Heterocycles in Bio-organic Chemistry

Heterocycles in Bio-organic Chemistry

Edited by
J. Bergman
The Royal Institute of Technology, Stockholm, Sweden

H. C. Van der Plas
Agricultural University de Dreyen, Wageningen, The Netherlands

M. Simonyi
*Central Research Institute for Chemistry,
Hungarian Academy of Sciences*

British Library Cataloguing in Publication Data
FECHEM Conference on Heterocycles in Bio-organic Chemistry
 (6th: 1990: Solbacka: Sweden)
  Heterocycles in bio-organic chemistry.
  1. Heterocyclic compounds
  I. Title II. Bergman, J. III. Plas, H. C. Van der (Henrik
  Cornelis) IV. Simonyi, Miklos V. Series
  547.59

ISBN 0-85186-877-0

The Proceedings of the Federation of European Chemical Societies
Sixth FECHEM Conference on Heterocycles in Bio-organic Chemistry,
held 27th—30th May 1990 in Solbacka, Sweden.

© The Royal Society of Chemistry 1991

All Rights Reserved
No part of this book may be reproduced or transmitted in any form
or by any means — graphic, electronic, including photocopying,
recording, taping or information storage and retrieval systems —
without written permission from The Royal Society of Chemistry

Published by The Royal Society of Chemistry,
Thomas Graham House, Science Park, Cambridge
CB4 4WF

Printed and bound in Great Britain by Bookcraft (Bath) Ltd.

Preface

These proceedings contain lectures, presented by invited speakers at the 6th FECHEM Conference on "Heterocycles in Bio-Organic Chemistry", held at Solbacka (Sweden, May 1990) and in addition some updated versions of selected lectures from the 5th Symposium, presented at Bechyne (Czecho-Slovakia, May 1988). In the lectures the most recent information is presented concerning the synthesis, physical and pharmacological properties of important bioheterocycles (tetrahydropyran containing natural products, benzodiazepinones, indole and isoquinoline alkaloids, Nuphar alkaloids, as well as stereochemical effects in synthesis and conformational studies).

The editors of the proceedings believe that in this book a reasonable coverage is given of today´s work on the role of heterocyclic compounds in bio-organic chemistry. Hopefully the book will contribute to a better understanding of the mechanism of action of bioheterocycles in bioprocesses.

At this point we would also like to acknowledge the financial support for the meeting from the following organizations and companies:
The Swedish Board of Natural Sciences, The Swedish Academy of Sciences, Astra, Pharmacia, Kabi, Draco, Hässle, The Swedish Tobacco Co, Glaxo, Schering and Shell.

The Editors

Contents

Synthesis of Tetrahydropyran-containing   1
Natural Products
By S.V. Ley

Selectivity and Flexibility in Biogenetically   18
Patterned Synthesis of Heterocyclic Compounds
By E. Winterfeldt

Chemo-enzymatic Approach to Some Indole   28
Alkaloids
By B. Danieli, G. Lesma, M. Mauro, and G. Palmisano

Sulfur-substituted Tryptophan - The Core of   54
Phallotoxins, Virotoxins, and Amatoxins
By H. Faulstich

Chemistry and Biology of Ascorbigens   68
*By M.N. Preobrazhenskaya, A.M. Korolev,
I.L. Plikhtyak, I.V. Yartseva, S.A. Efimov,
E.I. Lazhko, and L.G. Aleksandrova*

Some New Data on Structure, Properties, and   87
Mechanism of Action of 1,2-Dihydro-3H-1,4-
benzodiazepin-2-ones
By S.A. Andronati and T.A. Voronina

Design, Synthesis, and Evaluation of Functional   103
Analogs of CC-1065
By D.L. Boger

Stereochemical Effects During Synthesis of New   130
Heterocycles *via* Intramolecular Dipolar
Cycloadditions
By A. Hassner

New Reactions of Partially Hydrogenated   144
Ideno[1,2-*b*]-pyridines
By G. Duburs, D. Muceniece, and V. Lusis

Synthetic Studies on Biologically Active   154
Isoquinoline Alkaloids
By L. Castedo, D. Domínguez, and E. Guitián

New Synthetic Approaches to Nuphar Alkaloids   180
By J. Nowacki and J.T. Wrobél

Antiallergic 4H-Pyrido[1,2-a]pyrimidin-4-ones: 187
Synthesis and Structure Activity Relationship
By I. Hermecz

Heterocycles Designed for the Modelling of the   219
Zinc Active Site of Liver Alcohol Dehydrogenase
By R.M. Kellogg and B. Kaptein

Chemical Topology: Interlacing and Knotting   223
Molecular Threads on Transition Metals
By J.-P. Sauvage

Subject Index   229

Synthesis of Tetrahydropyran-containing Natural Products

S. V. Ley

DEPARTMENT OF CHEMISTRY, IMPERIAL COLLEGE OF SCIENCE, TECHNOLOGY AND MEDICINE, LONDON SW7 2AY, UK

Biologically active natural products have provided the stimulus for organic synthesis for many years. Partly as a result of these efforts much has been discovered in terms of strategies, new reactions, and reagents that can be applied to many other problems. Despite these enormous advances there are still bond constructions, stereoselective reactions and functional group selectivities that are impossible to control by current methods. Therefore there is a need to continue to advance the art of synthesis, especially if chemists are to make an impact at understanding biological processes at a molecular level and to work on Nature's biopolymers.

Industry also requires constant synthetic advances to be made in its effort to prepare novel materials with selective properties. Moreover, these developments must be made in an increasingly resource and environmentally conscious age.

Our group has been particularly interested in the synthesis and properties of several biologically active natural products (Scheme 1), all of which contain hydropyran or other cyclic ether components. On more careful inspection it should be noticed that all these cyclic ether arrays additionally contain carbon-carbon bonds at the centre adjacent to the oxygen atoms. Methods for forming these bonds from readily available starting points would therefore be highly desirable, especially as current procedures are limited.

Thus we have been developing the use of the 2-benzenesulphonyl substituent as a potential solution to these problems[1,2,3,4]. We reasoned that 2-benzenesulphonyl cyclic ethers could easily be prepared from lactols, lactol ethers or dihydropyrans and that we could exploit the dual role of the group by direct nucleophilic displacement[4] or alternatively by deprotonation and reaction with electrophiles (Scheme 2). In principle these methods would be very versatile and could readily be adapted to provide spiroacetal structures found in the commercially important avermectins and milbemycins[5,6]. It turns out that these sulphones are

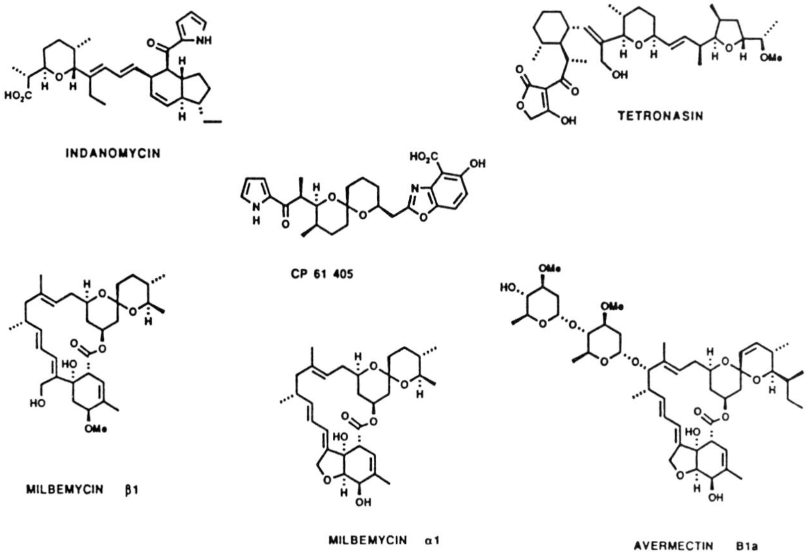
INDANOMYCIN
TETRONASIN
CP 61 405
MILBEMYCIN β1
MILBEMYCIN α1
AVERMECTIN B1a

Scheme 1

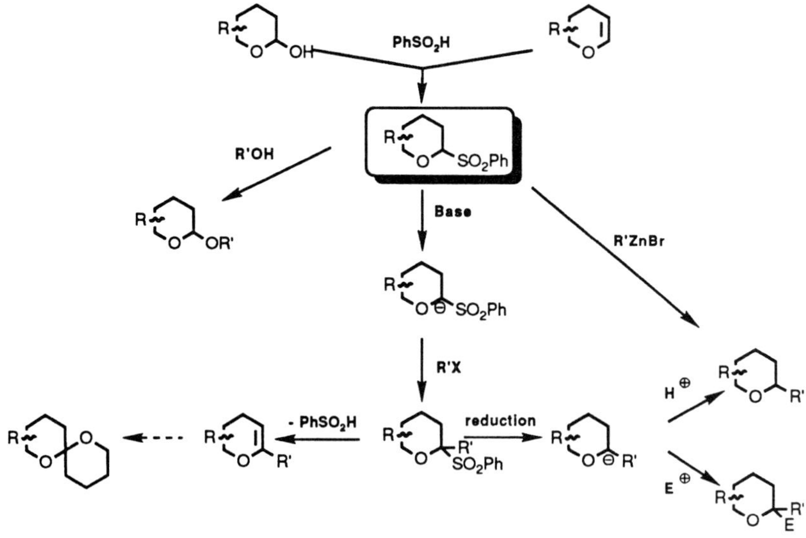
PhSO2H
R'OH
Base
R'ZnBr
R'X
- PhSO2H
reduction
H⊕
E⊕

Scheme 2

readily prepared, are relatively stable, often crystalline, and do undergo a very wide range of useful reactions. For example, under extremely mild conditions, involving magnesium bromide and sodium hydrogen carbonate, they react with alcohols to afford the corresponding acetals in excellent yields (Scheme 3). The mild conditions are tolerant of acid sensitive groups and reaction occurs even with hindered alcohols. If carbohydrate derived sulphones are used, the route constitutes a new glycosidation procedure.

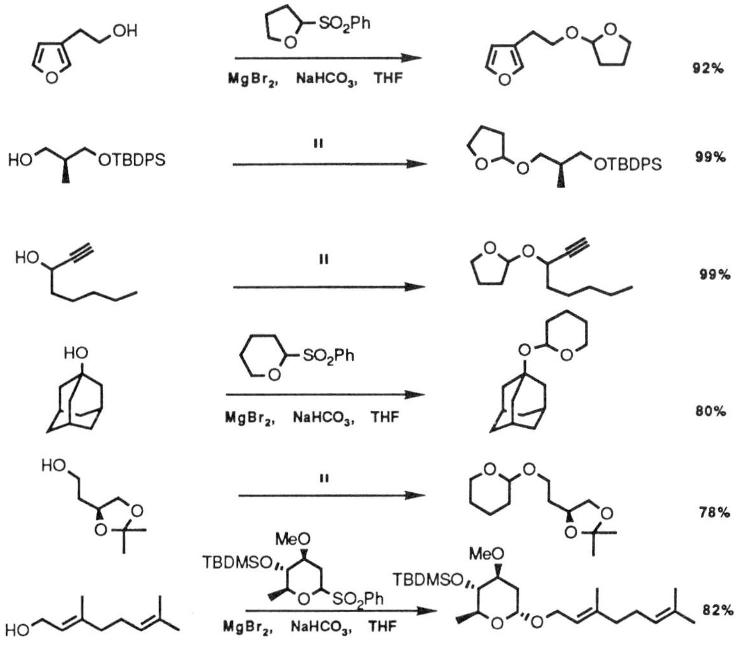

Scheme 3

Of more synthetic importance, however, is that direct nucleophilic displacement of the sulphone substituent with carbon nucleophiles is possible. Using organometallic reagents generated from Grignard reagents and zinc bromide, aryl, heteroaryl, alkenyl and alkynyl undergo substitution readily at room temperature (Scheme 4,5). More recently, alkyl organometallic reagents have been shown to effect similar displacement in dichloromethane. Nucleophilic substitution with silylenol ethers, allylsilane and trimethylsilyl cyanide in the presence of aluminium trichloride at -78°C occurs readily (Scheme 5). In reactions where a further ring substituent is present one obtains preferentially the trans-product irrespective of the starting sulphone geometry. Most notable from the examples in Scheme 4 is the stereoselective preparation of the bis-trimethoxyaryl furan recently declared by Merck as a Platelet Activating Factor antagonist (L-652,731)[7].

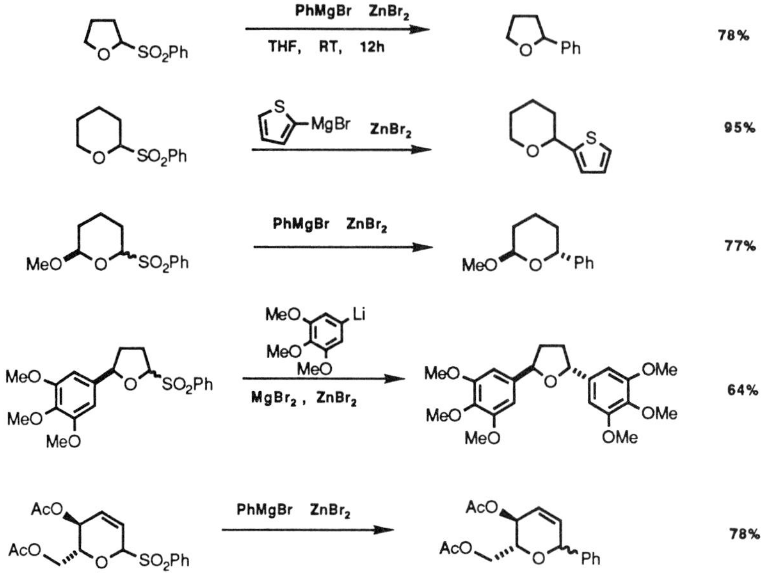

Scheme 4

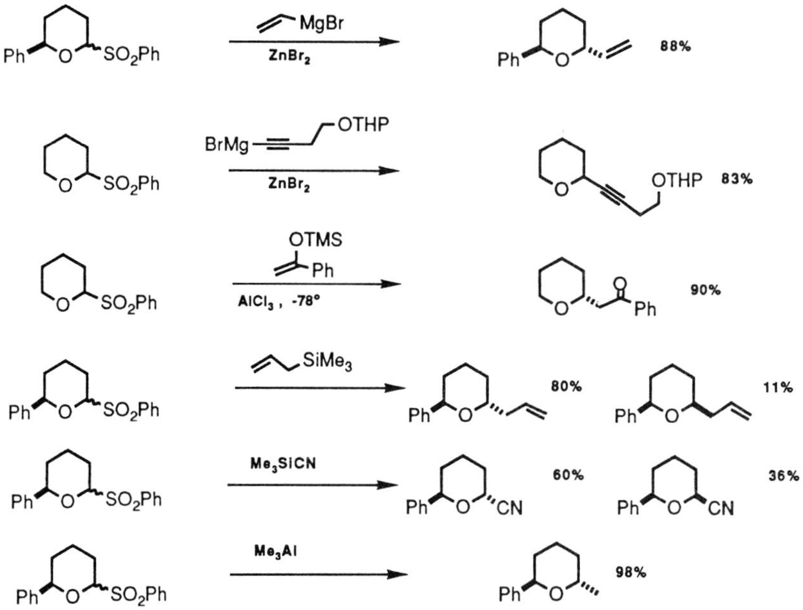

Scheme 5

As an illustration of the alternative properties of these cyclic ether sulphones, deprotonation and reaction of the resulting anions with substituted halides affords an efficient and concise preparation of spiroacetals (Scheme 6). For more elaborate spiroacetals containing hydroxyl groups and substitution patterns common to the milbemycins and avermectins, reaction of the sulphone carbanions with epoxides proves to be an excellent procedure (Scheme 7). The examples show the versatility of the method in that one can define the product substitution by appropriate choice of the coupling partners. The remaining problem to be solved in the avermectin model studies was to define a sequence which would also establish a double bond in the resulting spiroacetal. This was readily accomplished using a selenium electrophile to intercept the intermediate enol ether in these coupling reactions and hence incorporate a readily eliminated group in the spiroacetal[8] (Scheme 8). Much of the chemistry which has been described above for replacement of the benzenesulphonyl group at the anomeric position in cyclic ethers can also be translated into the cyclic amine series of piperidines and pyrrolidines[4].

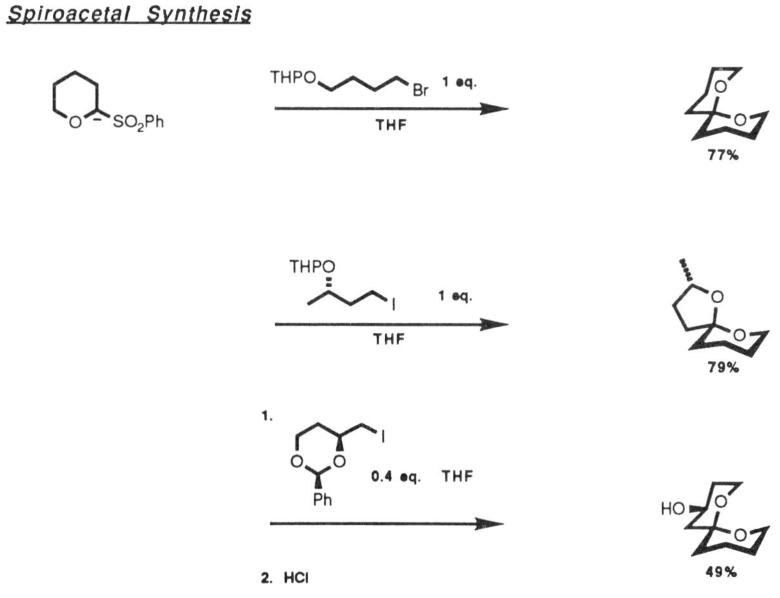

Spiroacetal Synthesis

Scheme 6

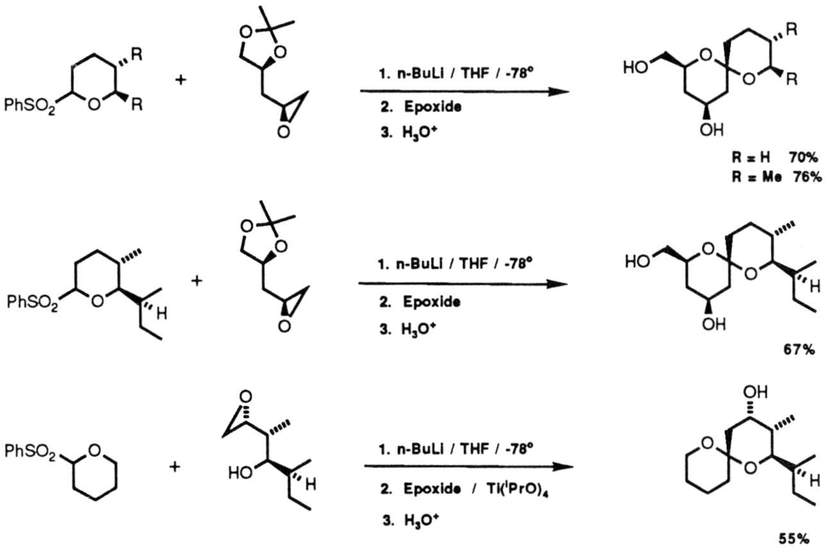

Scheme 7

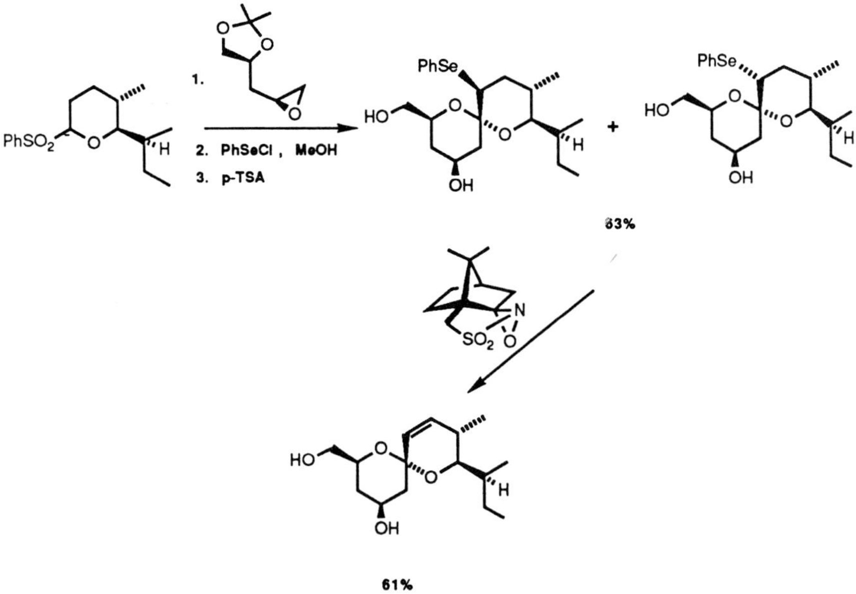

Scheme 8

In an effort to develop novel methodology to prepare the initial lactols (or lactones) used in this sulphone based chemistry we have been exploring the use of π-allyltricarbonyliron lactone complexes[9]. As these complexes are available from a variety of starting materials we argued that these easily handled materials are potentially very versatile precursors for lactones. Depending upon the decomplexation method and the substitution used, one could obtain lactones of different ring sizes (Scheme 9). Indeed, this approach can be used to selectively afford small ring β-lactones upon treatment with ceric ammonium nitrate[10], (Scheme 10) or, under exhaustive carbonylation condition, δ-lactones (see later). This chemistry can be used to prepare more elaborate natural product β-lactones and β-lactams[11] by choice of the appropriate substitution patterns.

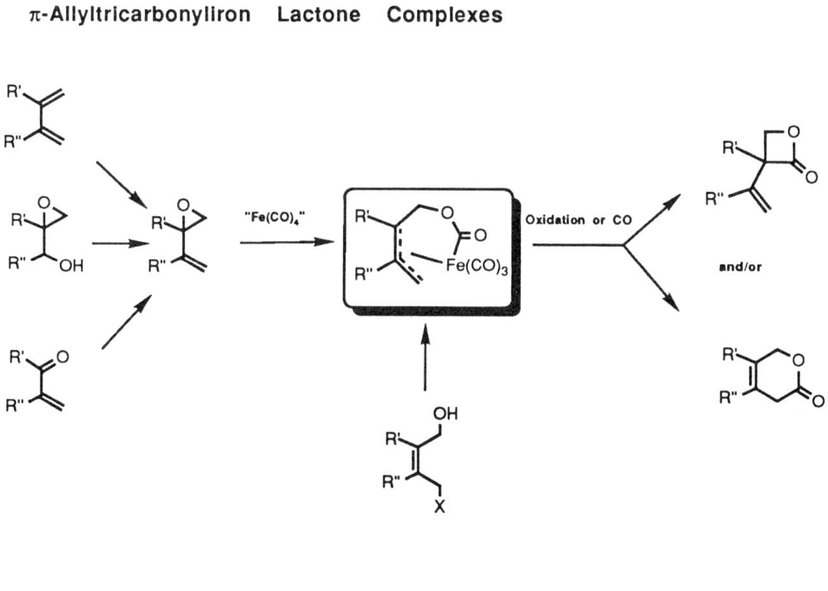

π-Allyltricarbonyliron Lactone Complexes

Scheme 9

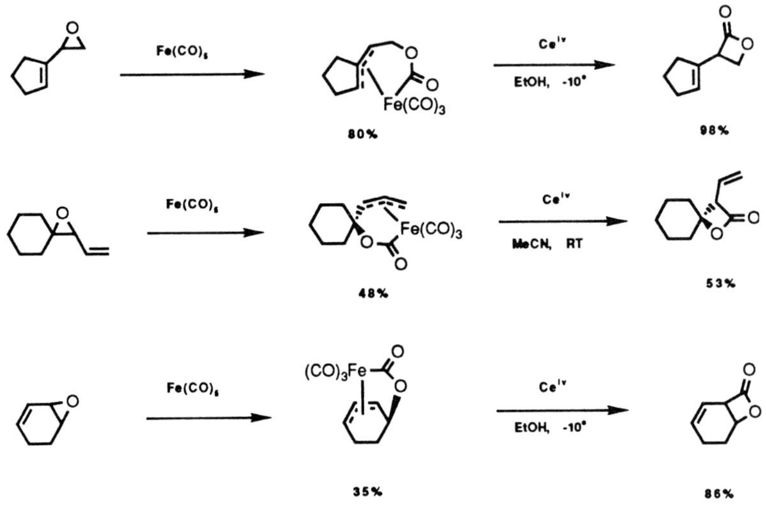

Scheme 10

With the above new methods firmly established, we sought to apply these procedures to more challenging target molecules. The ionophore antibiotic CP 61405, recently described by Pfizer[12], provides an excellent opportunity since it contains a bis-hydropyran spiroacetal group and a pyrrole carbonyl unit where we could additionally use chemistry developed by our group[13]. In Schemes 11 and 12 one can see the outcome of these studies and the progress towards CP 61405. Apart from the iron carbonyl, sulphone and 2-lithio-SEM-pyrrole methods one should also notice the application of tetra-n-propylammonium perruthenate (TPAP)[14] as a new catalytic, room temperature oxidant for alcohols, which we have developed.

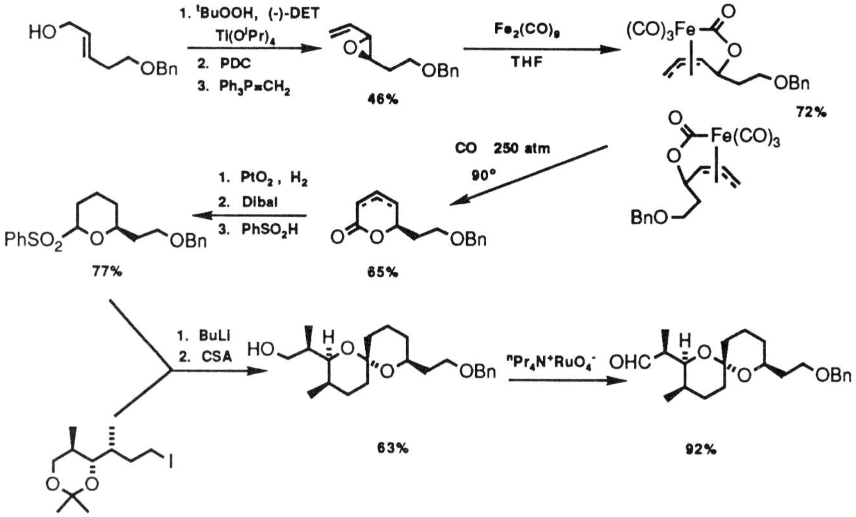

Scheme 11

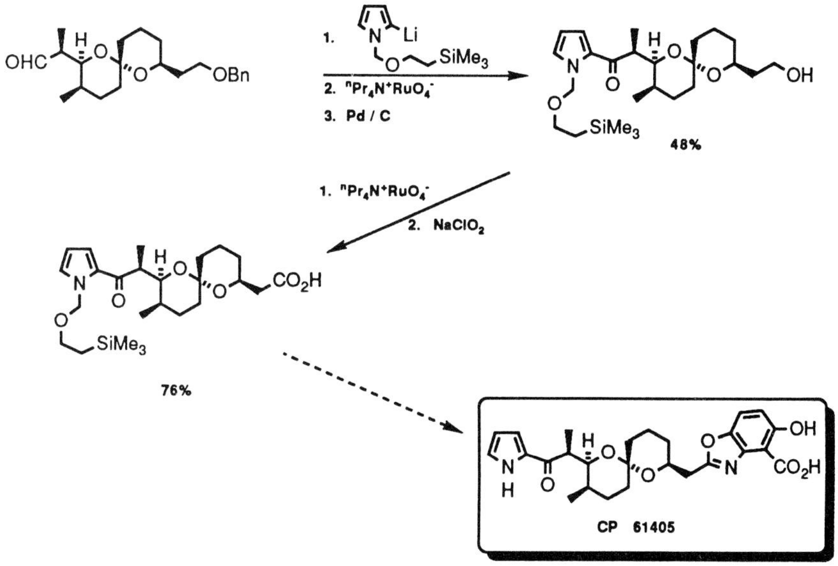

Scheme 12

Finally, using a combination of the above methods, we have completed the synthesis of the potent macrolide antiparasitic agent avermectin B1a[15]. The overall strategy for this synthesis broadly follows that previously shown by us to be successful for the related milbemycins. The highly convergent approach is shown in outline in Scheme 13 and the reader is referred to the publications for a discussion of the choice of coupling fragments in this scheme. Also we will not report here all the details of the preparation of the individual components in this scheme, rather we will concentrate on the new methods and overall philosophy of the approach.

Avermectin B1a

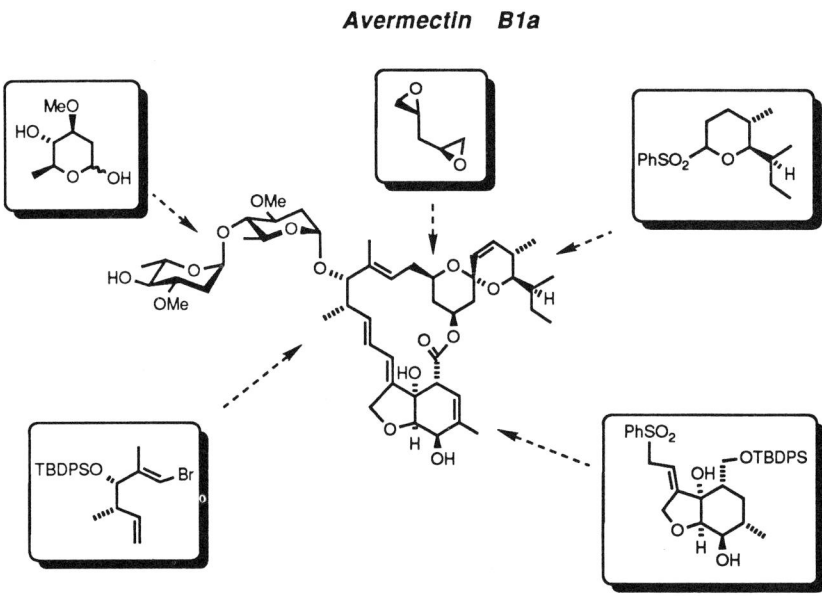

Scheme 13

Firstly, construction of the necessary carbohydrate hydropyran fragment, oleandrose, once again makes use of tricarbonyliron lactone technology (Scheme 14,15), as does preparation of the unsaturated-spiroacetal portion. Coupling of the oleandrose to the disaccharide employs the use of carbonyldiimidazole in the presence of Lewis acid ($ZnBr_2$), again a new glycosidation procedure which we developed specifically for this synthesis[16]. The choice of the thiocarbonyl imidazoyl in the disaccharide likewise was developed to effect final coupling to the avermectin B1a aglycone (Scheme 15).

Oleandrose Synthesis

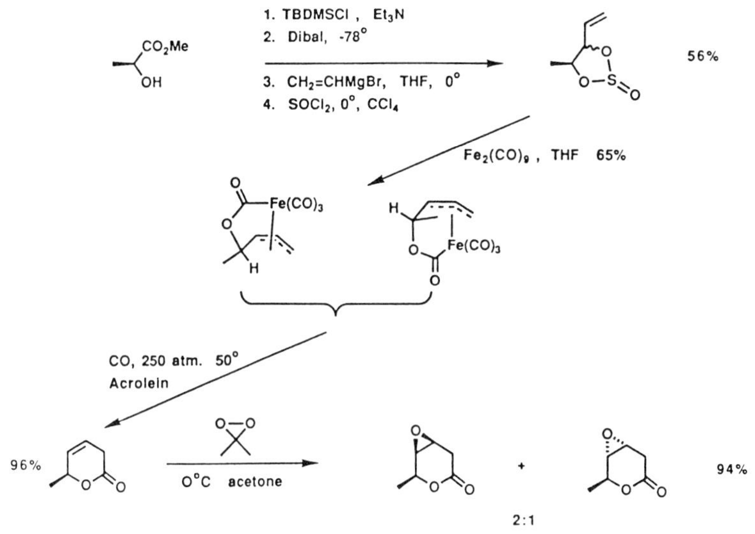

Scheme 14

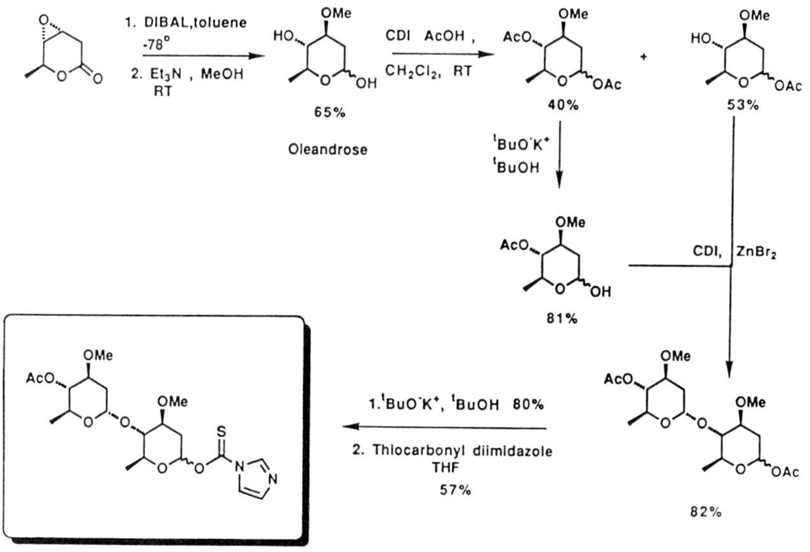

Scheme 15

For the preparation of the northern hemisphere fragment, use was made of sulphone stabilized anion coupling of epoxides and introduction of the required 22,23-double bond *via* the selenium approach discussed earlier (Scheme 16). This synthesis of the northern hemisphere fragment is extremely short compared with other routes. Not only does it feature the new chemistry for establishing the unsaturated spiroacetal position, it shows the importance of the diepoxide as a bis-electrophile and as a key unit for setting up the C-17 and C-19 stereogenic centres common to all avermectins and milbemycins. It should also be noticed in this scheme and in a later scheme (18) that the TPAP reagent played an important role as an oxidant.

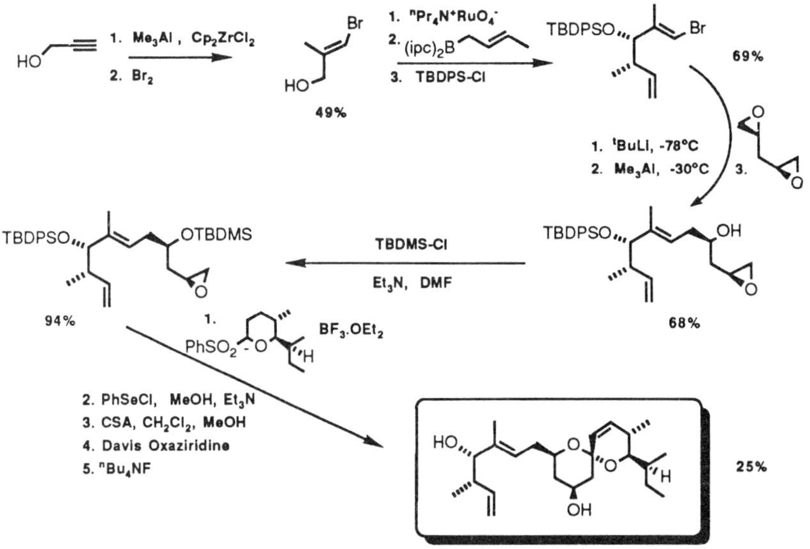

Scheme 16

In the remaining Schemes 17, 18, 19, 20 and 21 the completed synthesis is reported. Most noteworthy is bold decision to remove all alcohol protecting groups at an early stage in the synthesis. This might be considered by some to be foolish; however, we believed that by appropriate choice of reagents and conditions we could beneficially exploit the reactivity differences between the various hydroxyl groups. In this way we were able to complete the preparation of the aglycone from the initially coupled materials *without the need for hydroxyl group protection*. In order to set up these studies the synthetic northern hemisphere fragment was converted to an aldehyde suitable for coupling with a trianion derived from the C1-C10 southern allylic sulphone unit (Scheme 17). Following reductive

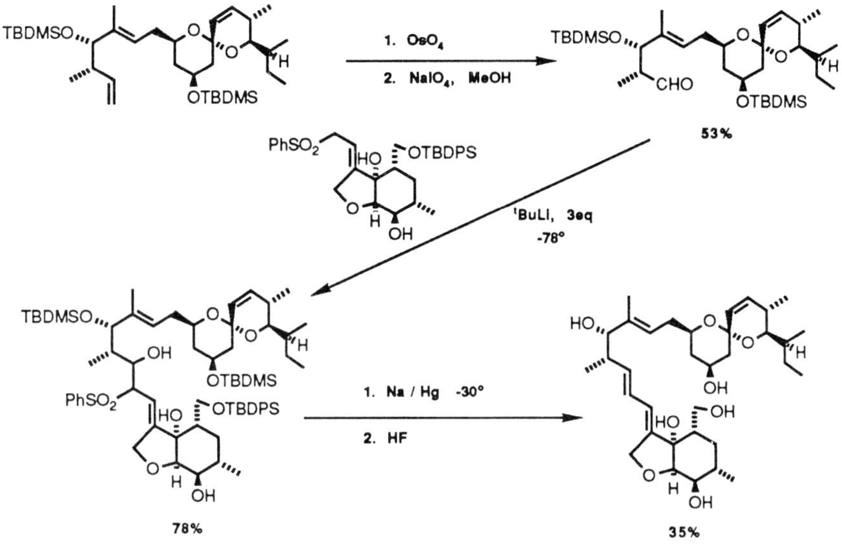

Scheme 17

desulphonylation and deprotection with HF in pyridine a pentol was obtained on which several selective hydroxyl group transformations were achieved. These include the selective oxidation of the primary alcohol followed by macrolactonisation. In order to introduce C3, C4 unsaturation, it was necessary to effect selective oxidation of the C-5 hydroxyl group. This was achieved by the stoichiometric application of TPAP reagent at 0°C which gave the product without substantial oxidation of the C-13 allylic hydroxyl group. This result was anticipated on steric grounds as the C-13 position is very inaccessible (Scheme 18). Final introduction of the C3-C4 double bond *via* selenenylation of the silyl enol ether of the C-5 carbonyl group and oxidation and *syn*-elimination afforded the aglycone (Scheme 19, 20). At this time we also see some elimination to give the *exo*-methylene product (29% yield). Work is in hand to vary the conditions to favour the desired endo-product (Scheme 20). The C-13 glycosidation of the aglycone required initial protection of the C-5 hydroxyl group as its acetate. Coupling of the fragments proceeded well however, giving a reliable 64% yield of anomerically pure product! Removal of the acetates at C5 and C4" was straightforward using super hydride to give avermectin B1a (Scheme 21).

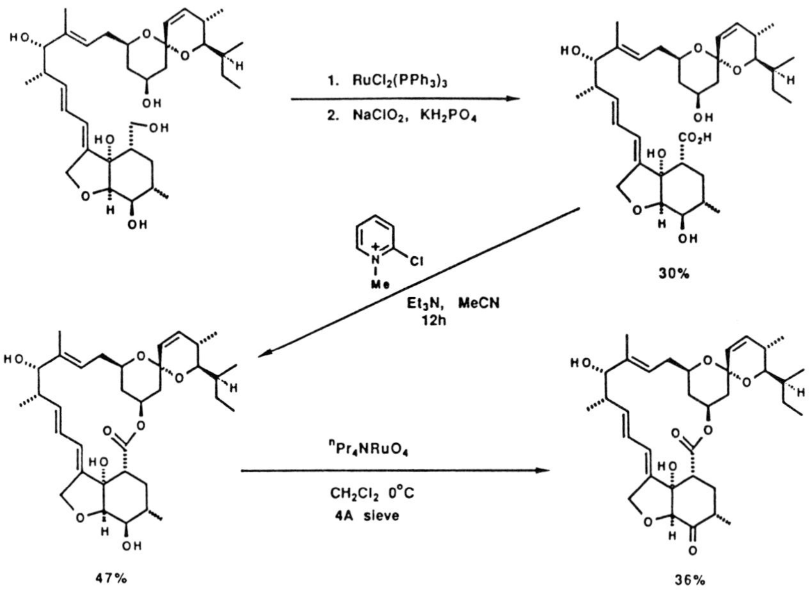

Scheme 18

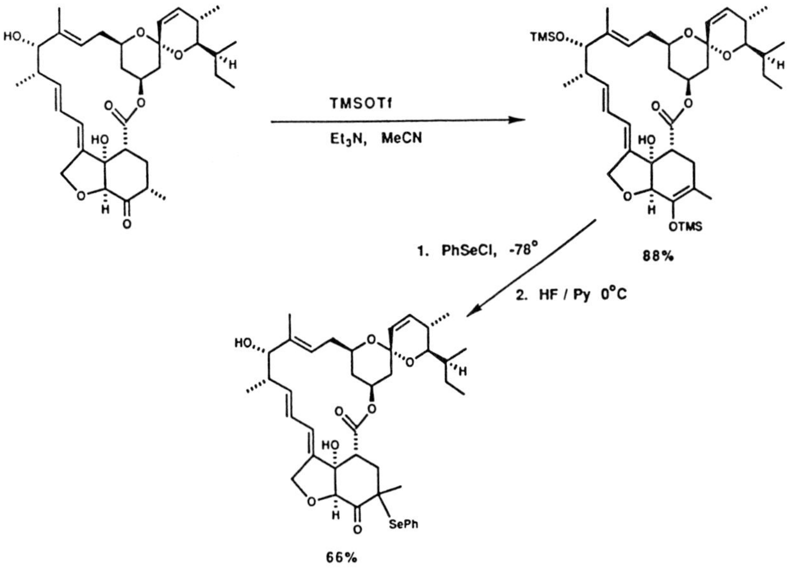

Scheme 19

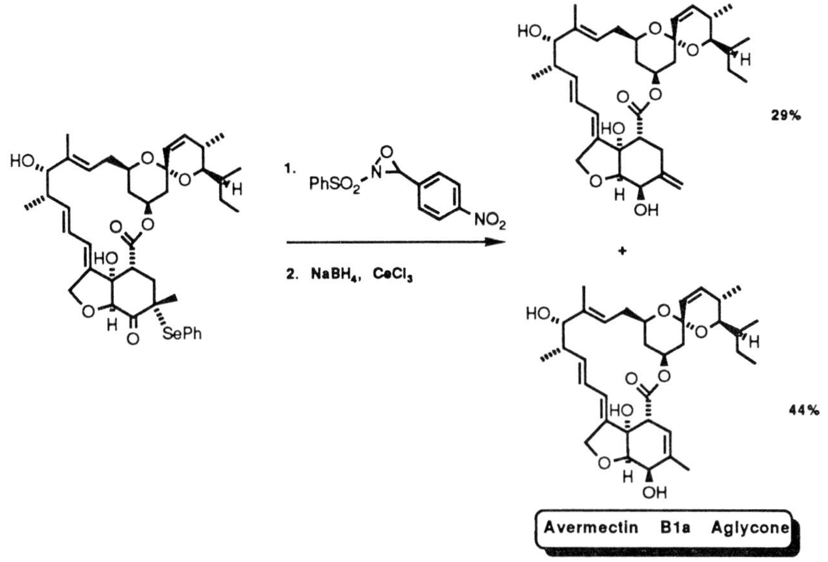

Synthesis of Tetrahydropyran-containing Natural Products

Scheme 20

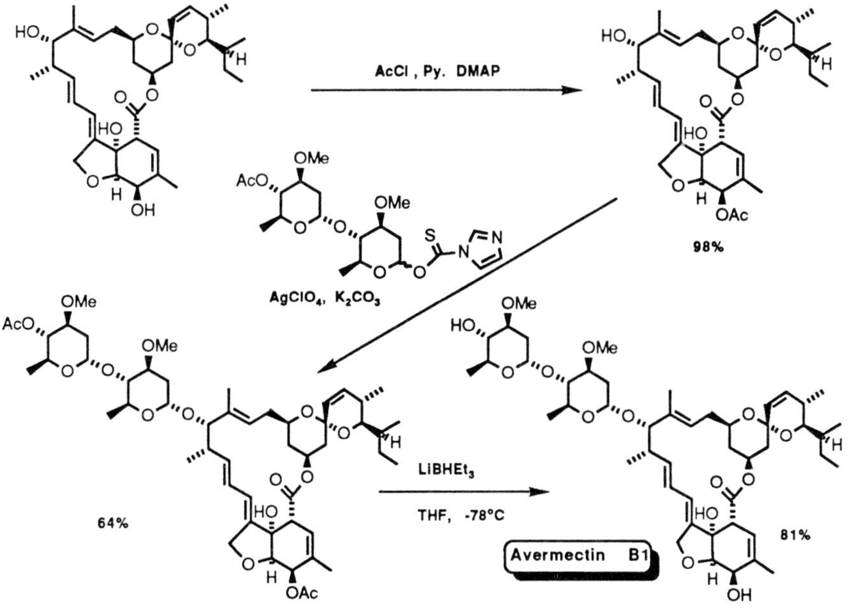

Scheme 21

In conclusion we have shown that by studying the synthesis of complex natural products one can make discoveries of new reagents and methods that should find application in a wider range of chemical programmes and in particular those of relevance to the pharmaceutical industry.

Acknowledgement: I thank all my excellent and energetic collaborators in these studies and the SERC, Pfizer Central Research, Merck Sharp and Dohme and ICI for financial support.

References:

1. S.V. Ley, B. Lygo, F. Sternfeld and A. Wonnacott, *Tetrahedron*, **1986**, *42*, 4333.
2. D.S. Brown, S.V. Ley and S. Vile, *Tetrahedron Lett.*, **1988**, *29*, 4873.
3. a) D.S. Brown and S.V. Ley, *Tetrahedron Lett.*, **1988**, *29* 4869; b) D.S. Brown, S.V. Ley and M. Bruno, *Heterocycles*, **1989**, *28*, Spec. Issue, No.2, 773; c) D.S. Brown, R.J. Davenport and S.V. Ley, *Tetrahedron*, **1989**, *45*, 4293.
4. D.S. Brown, T. Hansson and S.V. Ley, *Synlett*, **1990**, 48.
5. S.V. Ley, N.J. Anthony, A. Armstrong, M.G. Brasca, T. Clarke, C. Greck, P. Grice, A.B. Jones, B. Lygo, A. Madin, R.N. Sheppard, A.M.Z. Slawin and D.J. Williams, *Tetrahedron*, **1989**, *45*, 7161.
6. C. Greck, P. Grice, S.V. Ley, and A. Wonnacott, *Tetrahedron Lett.*, **1986**, *27*, 5277.
7. T. Bifu, N.F. Gamble, T. Doebber, S-B Hwang, T-Y. Shen. J. Synder, J.P. Springer and R. Stevenson, *J. Med. Chem.* **1986**, *29*, 1917.
8. D. Diez-Martin, P. Grice, H.C. Kolb, S.V. Ley and A. Madin, *Tetrahedron Lett.*, **1990**, 3445.
9. S.V. Ley, *Phil. Trans. R. Soc. Lond. A.* **1988**, *326*, 633.
10. G.D. Annis, S.V. Ley, C.R. Self, and R. Sivaramakrishnan, *J. Chem. Soc., Perkin Trans. 1*, **1981**, 270.
11. a) G.D. Annis, E.M. Hebblethwaite, S.T. Hodgson, D.M. Hollinshead, S.V. Ley, *J. Chem. Soc., Perkin Trans. I*, **1983**, 2851; b) S.T. Hodgson, D.M. Hollinshead and S.V. Ley, *Tetrahedron*, **1985**, *41*, 5878.
12. U.S.P. 4547523/1985
13. M.P. Edwards, S.V. Ley, S.G. Lister and B.D. Palmer, *J. Org. Chem.*, **1984**, *49*, 5503; M.P. Edwards, A.M. Doherty, S.V. Ley and H.M. Organ, *Tetrahedron*, **1986**, *42*, 3723.
14. W.P. Griffith, S.V. Ley, G.P. Whitcombe and A.D. White, *J. Chem. Soc., Chem. Commun.*, **1987**, 1625; W.P. Griffith and S.V. Ley, Aldrichimica Acta, 1990, 13.
15. a) A. Armstrong and S.V. Ley, *Synlett*, **1990**, 323; B) D. Diez-Martin, P.

Grice, H.C. Kolb, S.V. Ley and A. Madin, *Synlett*, **1990**, 326; c) A. Armstrong, S.V. Ley, A. Madin and S. Mukherjee, *Synlett*, **1990**, 328; d) M.J. Ford, J.G. Knight, S.V. Ley, S. Vile, *Synlett*, **1990**, 331.
16. M.J. Ford and S.V. Ley, *Synlett*, **1990**, 255.

Selectivity and Flexibility in Biogenetically Patterned Synthesis of Heterocyclic Compounds

E. Winterfeldt

INSTITUTE FÜR ORGANISCHE CHEMIE, UNIVERSITÄT HANNOVER,
SCHNEIDERBERG 1B, D-3000 HANNOVER 1

In an earlier contribution from our laboratory a biogenetically oriented approach to **4** - the aglycone of staurosporine, **3**,[1,2] - was described (*Scheme 1*) which proved that a precursor like **2** could simply generate the target molecule **4** in a sequence of photocyclization and dehydrogenation.[3]

Scheme 1

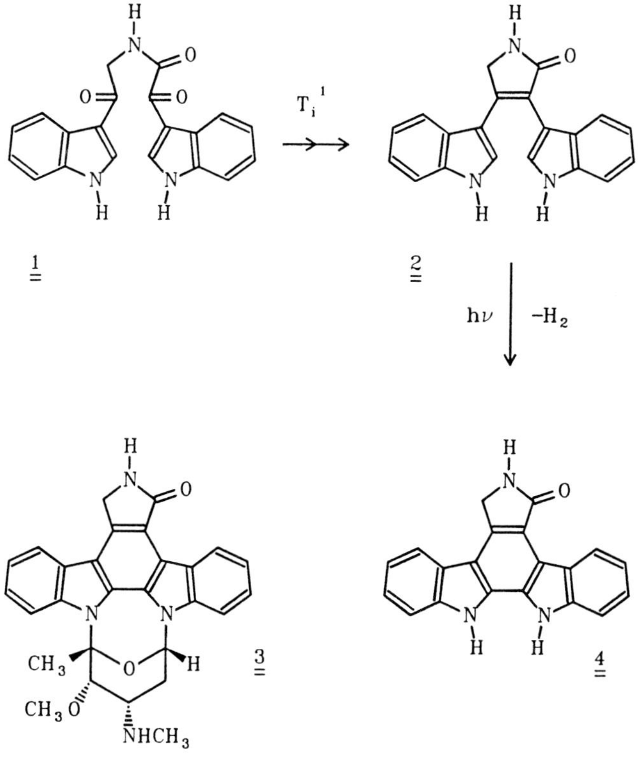

Although this does represent a quite straightforward access to the ring system, we were unable to insert the missing six-carbon link to generate the natural product itself. This demanded a change in strategy, as on the one hand, the preparation of the diketone **1** demands non substituted indole ring systems (SeO$_2$-oxidation), while on the other hand regioselective alkylations at stages **1** and **2** were not considered promising transformations.

The general decision was to use two different N-substituted indole building blocks and to merge them to an intermediate of type **2** without employing this time the SeO$_2$-oxidation. This would leave the option to at a later stage again use the very simple and efficient photocyclization to gene-rate the indolo-pyrrolocarbazole system.

Scheme 2

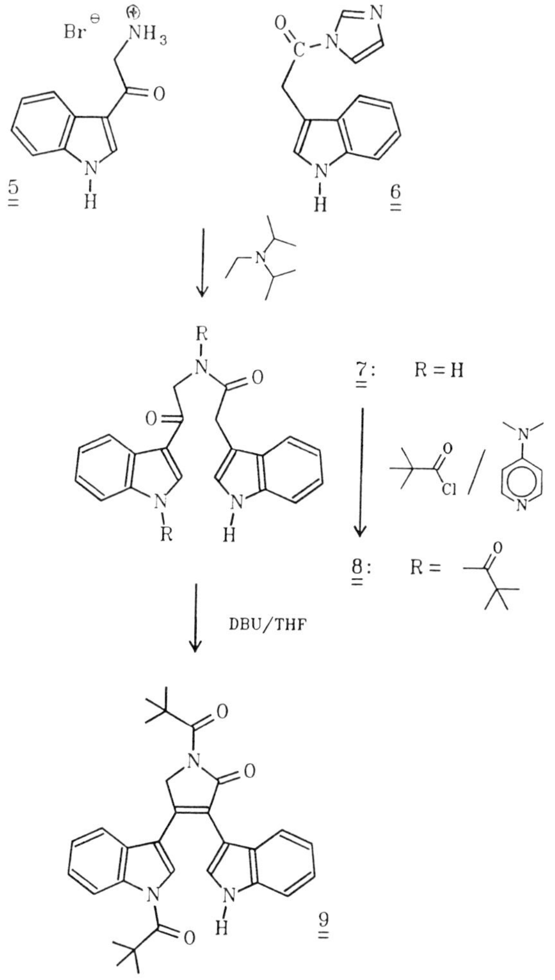

To test a very simple derivative we studied the cyclization of compound **8** which is easily obtained from amino ketone **5** and activated indolylacetic acid **6** followed by regioselective acylation (NH-acidity). Pivaloyl chloride was chosen to exclude additional acidic protons a to a carbonyl group

While very strong bases did not lead to any cyclization products, DBU in THF gave rise to the desired product **9** wanted, albeit only in 15% yield. The explanation for this result is not straightforward. While the pivaloyl residue on the amino ketone moiety should activate this very carbonyl group, owing to its bulkiness could also hinder the cyclization process. On the other hand the second still quite acidic indole proton might be the culprit.

To learn more about the role of the various NH groups in this trans-formation we prepared the monomethyl analogue **10a** of ketoamide **7** (see *Scheme 3*).

Scheme 3

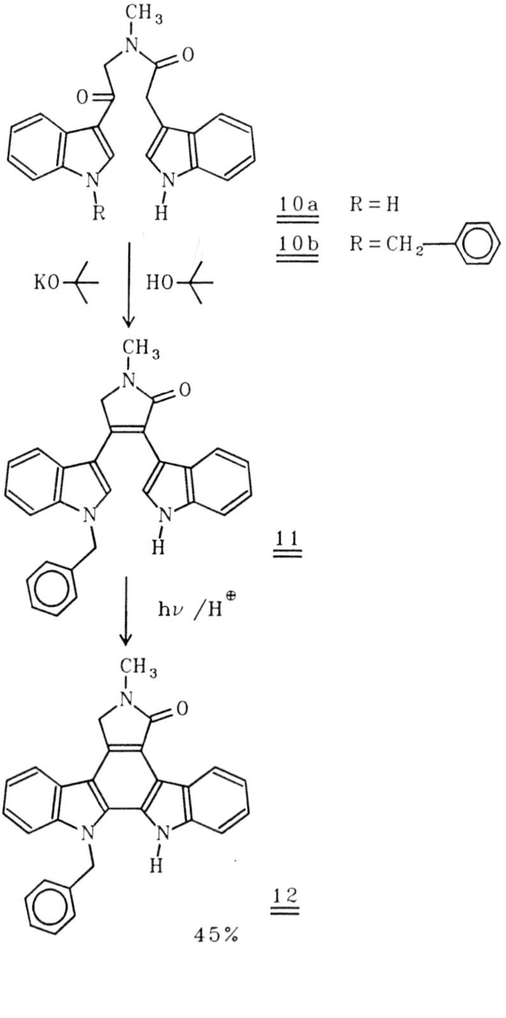

As one might expect **10a** proved to be reluctant to cyclization. Mono-alkylation with benzyl bromide gave the N-benzyl compound which afforded a respectable 83% of the cyclization product **11**. While in this case DBU and LDA in THF failed completely, equilibrating and transprotonation conditions (potassium tert-butylate in tert-butanol) proved to be the medium of choice, proving that one acidic centre may be tolerated in the cyclization process. Finally, in this series the successful photocyclization under acidic conditions (ethanol/TFA) has to be mentioned, proving that one substituent on the indole nitrogen does not block this process.

Scheme 4

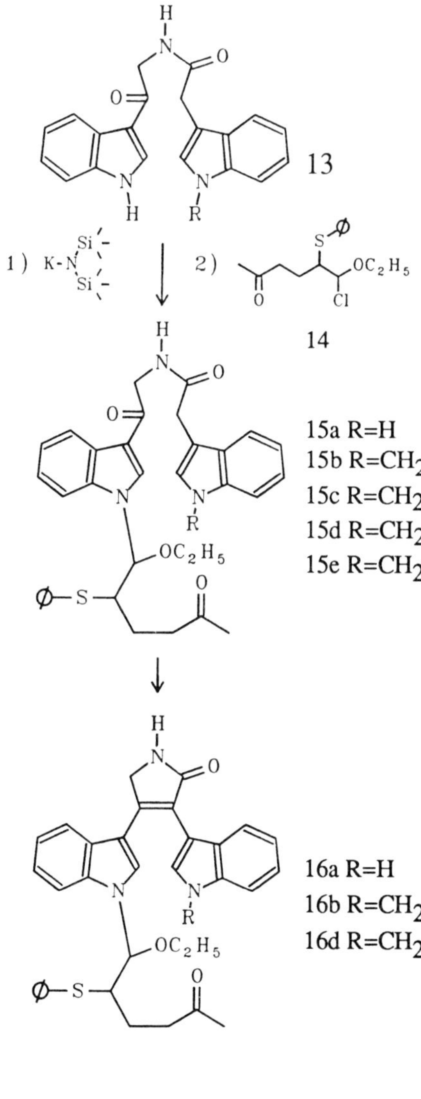

15a R=H
15b R=CH$_2$-Φ
15c R=CH$_2$-Φ-OCH$_3$
15d R=CH$_2$-OCH$_3$
15e R=CH$_2$CH$_2$-Si-

16a R=H   9%
16b R=CH$_2$-Φ   60%
16d R=CH$_2$OCH$_3$ 85%

As the natural product represents a non-substituted γ-lactam we decided to test intermediates with this substitution pattern to hopefully arrive at cyclization products with an NH group in this position. Being aware of the possibility for regioselective alkylations at stage **10** we started with various N-substituted indolyl acetic acid derivatives to prepare compounds of type **13** (*Scheme 4*).

As expected (two NH groups !) they themselves were unsuccessful in attempted cyclization processes. After alkylation with compound **14**, however, which is easily obtained from the corresponding enol ether by treatment with pentyl sulfenyl chloride, smooth cyclizations were observed. Again **15a** (two NH groups!) gave the lowest yield but with disubstituted precursors yields went up to more than eighty percent (R=MOM; R=SEM).

Scheme 5

Photocyclizations on **16b** were totally unsuccessful indicating too much sterical hindrance in the disubstituted case. After hydrolysis of the MOM ether a photocyclization product was obtained in low yield which was assigned structure **17** (MS).

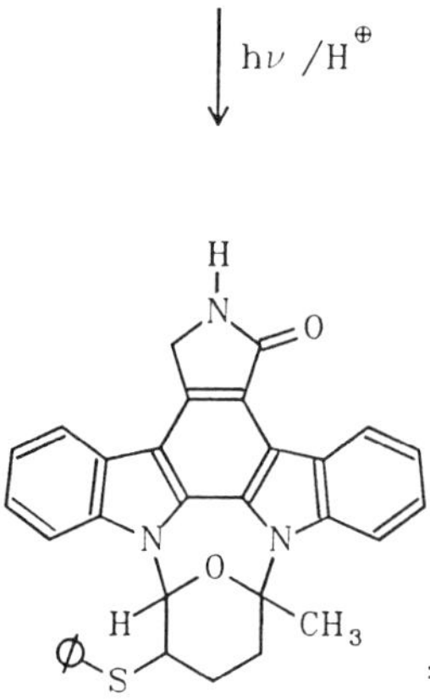

While in the first project we mainly dealt with problems of regioselec-tivity the second part will focus on a conformation dependent stereoselectivity in the spiropiperidine series. This investigation also started with a biomimetic synthetic approach. As Witkop[4] and coworkers had claimed a straight chain precursor for the spiropiperidine - as well as the pumiliotoxine type alkaloids - we prepared compound **18a** and were pleased to note that both ring systems indeed were formed on reductive amination (see *Scheme 6*).[5]

Scheme 6

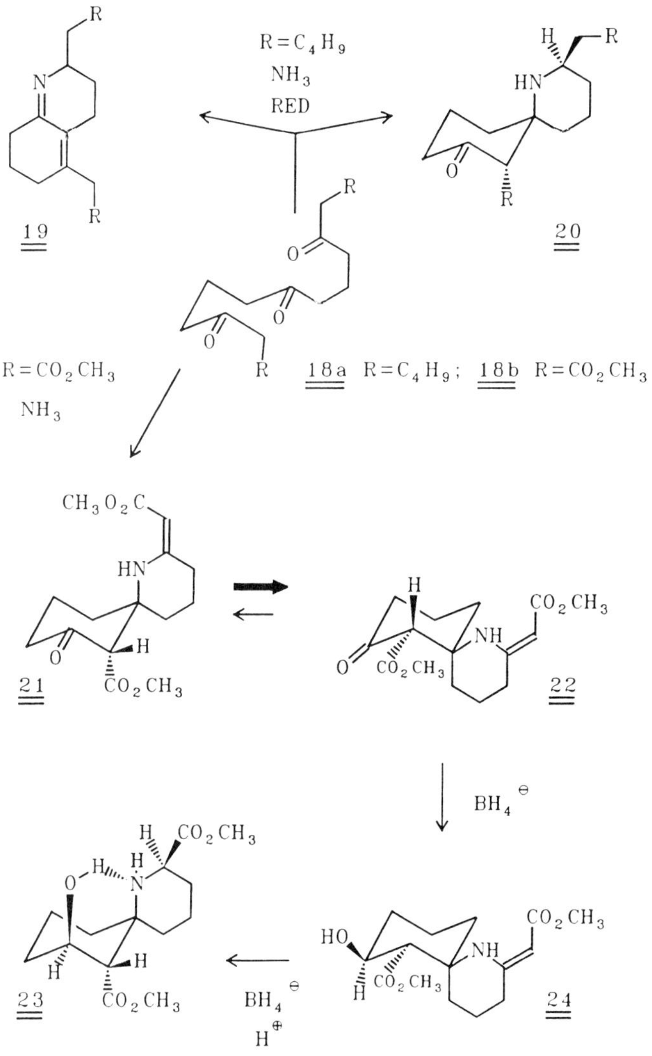

Although this is good news from the biogenetic point of view it is bad news for synthetic chemist as it means lack of regioselectivity. But there is another crushing information from the fact that the wrong configuration is installed at C_2 if one starts with an sp^{3}centre therefore the cyclization process (transition state of the intramolecular Michael addition).

To solve both problems at one stroke we changed over to the double β-ketoester **18b** (enhanced C-H acidity at the "outer" CH_2 groups, sp^2 at C_2) and were happy to observe smooth and regio-selective spiro cyclization to form **21** on simple treatment with ammonia. The subsequent borohydride reduction had to be done in two stages and gave rise again to the "wrong" configuration at C_2 (X-ray). Obviously the alcohol obtained from the β-ketoester populated conformation **24** (NMR data!) with the equatorial ester group hindering front side attack of the reducing reagent. After this reduction a strong hydrogen bridge forces the spiro system into conformation **23**.

As the size of substituents is known to strongly influence the enamine-imine equilibrium we were quite interested in the fate of some bulky esters we had prepared for enantioselective spiro cyclizations.

Scheme 7

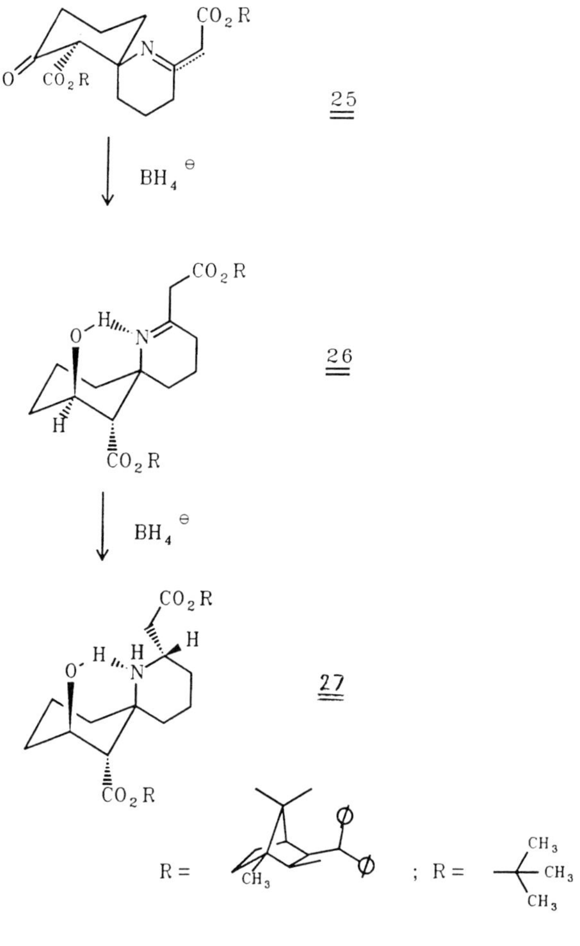

Out of a series of candidates ester **25** has proven to be by far the best as far as diastereoselectivity went. Only one cyclization product could be detec-ted with NMR techniques which indicated a selectivity of at least 96%. Besides yielding optically pure spiro compounds this special bulky diester offered exciting results from the borohydride reduction. As UV and IR spectra as well as NMR data (strongly reduced olefinic proton) had indicated the pre-vailing imine structure **25**, we were not surprised to note that this special ester could be taken to the tetrahydro compound without adding any acid at all (protonation of the enamine).

Scheme 8

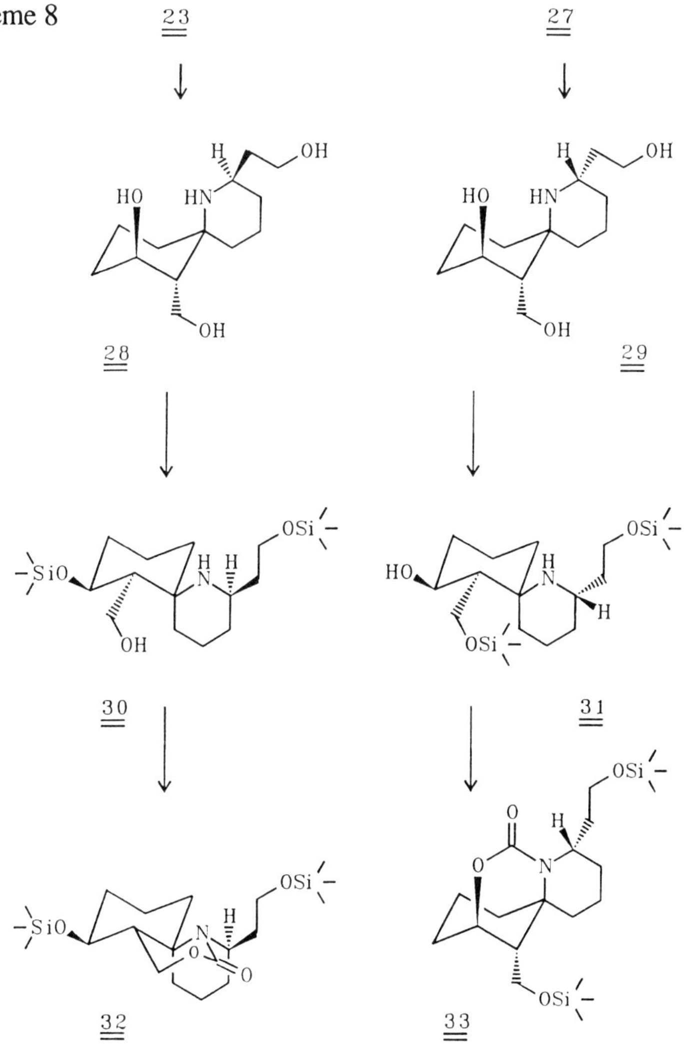

After reduction of the ketone, the basic imine again forces the spiro system into the hydrogen bridge stabilized conformation **26** which shows no front side hindering but additionally can complex the borohydride thus directing the hydride into b-attack to form the "natural" configuration at C_2. To make sure that this useful stereoselectivity is not induced by the special configuration of the auxiliary alcohol but by its sheer size we conducted the same set of experiments with a tert-butyl ester to note also formation of the natural β-configuration at C_2. Besides spectroscopic data to prove this event there is also an interesting chemical proof for the change in configuration. While the unnatural 2-α-configuration after hydride reduction and partial, selective silylation of the resulting triol generated bis-silyl ether **30,** the primary hydroxy group being hindered by the bulky C_2 substituent the "natural" triol afforded bis-silyl ether **31**. In the absence of special shielding of the second primary alcohol the "normal" sequence of reaction rates is restored. In accordance with this assignment the two monohydroxy com-pounds form very different cyclic urethanes with phosgene. Thus **32** obviously involves a primary hydroxy group (A,B-quartet, NMR) while **33** only shows one equatorial proton of this type (as configuration and conformation demand). These results prove that by manipulating the enamineimine equilib-rium with the help of the size of substituents one can direct the reduction and reach both configurations at C_2 of the spiropiperidine system, thus arriving at **27** with the correct absolute configuration for a large group of alkaloids.

REFERENCES

1. a) S. Omura, Y. Iwai, A. Nakayawa, J. Awaya, T. Tsuchiya, Y. Takahashi, and R. Masuma, J. Antibiotics, (1977), **30**, 275.
   b) A. Furusaki, N. Hashiba, and T. Matsumoto, J. Chem. Soc.,Chem. Commun., (1978), 800.
   c) W. Steglich, B, Steffan, L. Kopanski, and G. Eckhardt, Angew. Chem. (1980), **92**, 463.
2. a) S.M. Weinreb, R.S. Gainor, and S.M. Gainor, Heterocycles, (1984), **21**, 309.

- b) R.P. Joyce, J.A. Gainor, and S.M. Weinreb, J. Org. Chem. (1987), **52**, 1177.
- c) P.D. Magnus, C. Exon, and N.L. Sear, Tetrahedron, (1983), **39**, 3725.
- d) P.D. Magnus and N.L. Sear, Tetrahedron, (1984), **40**, 2795.
- e) B. Pelcman, Diss. Stockholm, (1988).
- f) J. Bergman and B. Pelcman, Tetrahedron Lett., (1987), **28**, 4441.
3. a) B. Sarstedt and E. Winterfeldt, Heterocycles, (1983), **20**, 469.
   b) B. Sarstedt, Diss. Hannover, (1982).
4. B. Witkop, Experientia (1971), **27**, 1121.
5. a) M. Glanzmann, C. Karalai, B. Ostersehlt, U. Schön, C. Frese, and E. Winterfeldt, Tetrahedron, (1982), **38**, 2805.
   b) E. Winterfeldt, Bull. Soc. Chim. Belg., (1988), **97**, 705.

Chemo-enzymatic Approach to Some Indole Alkaloids

B. Danieli, G. Lesma, M. Mauro, and G. Palmisano

DIPARTIMENTO DI CHIMICA ORGANICA E INDUSTRIALE, UNIVERSITA'
DEGLI STUDI DI MILANO, CENTRO CNR DI STUDIO PER LE SOSTANZE
ORGANICHE NATURALI, VIA VENEZIAN, 21 - 20133 MILANO, ITALY

1 INTRODUCTION

A fundamental step in the total synthesis of chiral natural compounds is the development of efficient methodologies for the production of enantiomerically pure starting materials. This strategy may involve the use of chiral templates derived from natural sources such as amino acids, carbohydrates, hydroxyacids and terpenes (the "chiron" approach) or asymmetric synthesis. An asymmetric synthesis can be performed chemically by the use of a stoichiometric or catalytic amount of a chiral reagent or by employing biological systems. In recent years there has been an increasing interest in the use of microorganisms or isolated enzymes, not only for the kinetic resolution of racemates and differentiation of enantiotopic faces in achiral precursors with prochiral centers, but also for the differentiation of compounds with σ-symmetry in their prochiral or meso form.
The interest in the resolution of the last mentioned compounds lies in the fact that symmetrical or latent symmetrical units are frequently present in molecules of natural origin.

The retrosynthetic analysis of these molecules (Figure 1) can be carried out in such a way to transform the target structures into a sequence of progressively simpler intermediates which ultimately lead to simple or commercially available symmetrical starting materials [1,2]. Such materials are represented in Figure 1 as diesters (dimethylesters or diacetates) as they are most commonly used. The synthetic plan is therefore based on the ability of enzymes to accomplish asymmetrization of σ-symmetrical diesters to lead to one intermediate with C_1-symmetry (methyl hydrogen dicarboxylates or dimethanol monoacetates), which can be converted into the target by non-enzymatic transformations expressly designed to respect the integrity of the stereogenic centers.

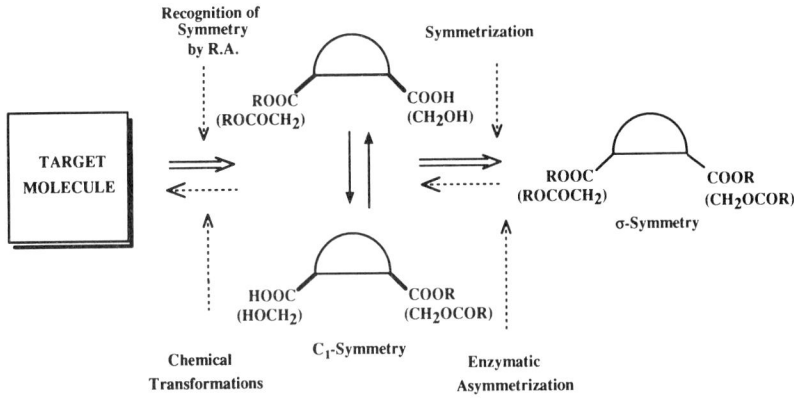

FIGURE 1

The recognition of a symmetrical unit in a target molecule can be difficult and can be a limitation of this Symmetrization-Asymmetrization approach. However, the possibility to perform a formal enantiomeric conversion of the two C_1 intermediates by suitable modulation of the reactivity of the two similar functional groups, gives the approach great flexibility.

In the following we will discuss synthetic approaches to some indole alkaloids in their optically active form, exploiting the concepts of Symmetrization-Asymmetrization.

2. SYNTHESIS OF ALKALOIDS FROM CIS-CYCLOHEX-4-ENE-1,2-DIMETHANOL DIACETATE.

First of all we will consider the monoterpenoid alkaloids belonging to Class I, exemplified in Scheme 1 by ajmalicine. It is well known that these compounds are biogenetically derived from tryptamine and secologanine and possess the intact carbon skeleton of the monoterpene precursor. Only a few of them have been obtained directly in an enantioconservative way from secologanine itself. The great majority has been synthesized by assembling the C_9-C_{10} unit through a suitable combination of a variety of fragments.

By reasoning along the lines of retrosynthetic symmetrization, the carbon framework of secologanine can be reconnected in two ways to a cyclohexene derivative:

i) reconnection through carbons 4 and 5 to a *cis*-1,2-disubstituted cyclohex-4-ene,

ii) reconnection through carbons 4 and 9 to a *trans*-cyclohexene.

By focusing the attention on the route to the *cis*-cyclohexene, minor adjustments of the number of carbon atoms and of the oxidation level first suggest a compound with C_1-symmetry and secondly a compound with σ-symmetry as starting materials.

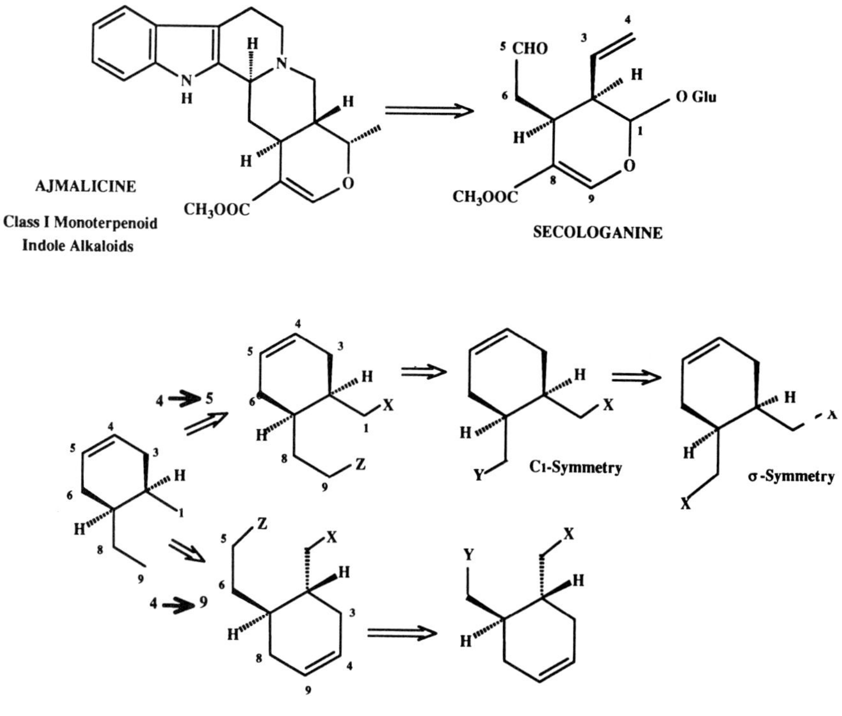

SCHEME 1

Symmetrical *cis*-1,2-disubstituted cyclohex-4-enes are very versatile synthons in the synthesis of natural products. In particular, racemic dimethyl *cis*-cyclo- hex-4-ene-1,2-dicarboxylate and *cis*-cyclohex- 4-ene-1,2-dimethanol diacetate (or equivalents) have been widely used in the preparation of terpenes, carbapenems, aminosugars and nucleoside antibiotics.

These cyclohexenes have also been extensively studied with respect to their asymmetrization with enzymes of the esterase and lipase type (Scheme 2).

Porcine liver esterase (PLE) shows a pro-*R* enantiotopic specificity for the hydrolysis of dimethylcarboxylate to afford the methyl hydrogen (+)-(1*S*,2*R*)-cyclohex-4-ene-1,2-dicarboxylate in good yield (94-99%) and optical purity (85-99% e.e.)[3].

With the cyclohexene dimethanol diacetate **1**, PLE gives the dimethanol monoacetate (+)-(1S,2R)-**2** with moderate yield (78%) and good e.e. (96%)[4], whereas pig pancreatic lipase (PPL) more efficiently produces the enantiomeric (-)-(1R,2S)-**2** (96% yield, 98% e.e.)[5].

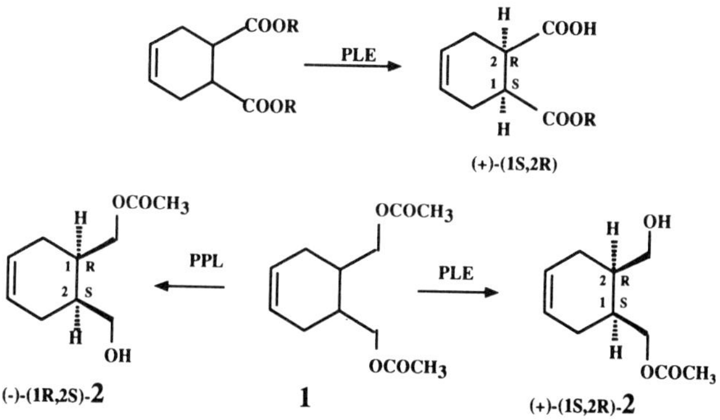

SCHEME 2

In Figure 2 the structures of the indole alkaloids (-)-antirhine and (-)-akagerine which we have synthesized starting from suitable chiral cyclohexene dimethanol monoacetate are shown. In addition we have also prepared (+)-meroquinene which, of course, is not an indole alkaloid, but a well known degradation product of the biogenetically related *Cinchona* bases.

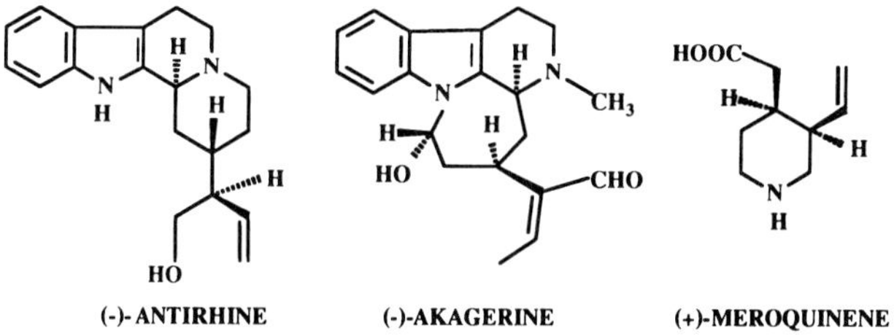

FIGURE 2

Antirhine, the major alkaloid of *Antirhea putaminosa* [6], is a unique variant of *Corynanthe* skeleton possessing a *cis* C/D ring junction and a less stable *anti*-relationship of the hydrogens at C-3 and C-15. A retrosynthetic analysis of antirhine, easily reveals that the non-indolic moiety could be derived from a *cis*-cyclohexene derivative (Scheme 3), thus generating the carbons at position 3 and 18 by cleavage of the double bond. Further disconnection leads to a lactone which ultimately is derived from the dimethanol monoacetate (-)-2 by elongation of the chain at C-1. This disconnection implies that the chirality at C-15 and C-20 of antirhine is determined by that at C-1 and C-2 of (-)-2, whereas that at C-3 might be induced during the cyclization step.

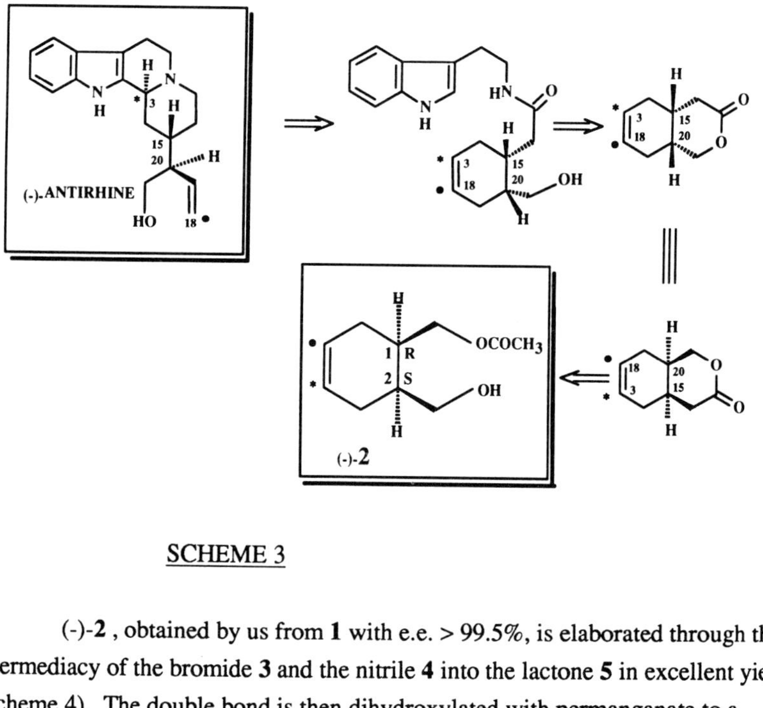

SCHEME 3

(-)-2, obtained by us from 1 with e.e. > 99.5%, is elaborated through the intermediacy of the bromide 3 and the nitrile 4 into the lactone 5 in excellent yield (Scheme 4). The double bond is then dihydroxylated with permanganate to a diastereomeric mixture of *cis*-diols. Condensation of the protected lactone 6 and tryptamine in refluxing n-BuOH gave the amide 7 (Scheme 5) which after deprotection and oxidation gave the stable dialdehyde 8.

Treatment of the dialdehyde 8 under carefully controlled conditions afforded the tetracyclic lactam 9 possessing a six-membered D ring as a 9:1 mixture of 3αH and 3βH epimers. One of the aldehyde groups is captured intramolecularly by the primary OH group to form a 2-hydroxytetrahydrofuran ring. Any traces of a lactam having a seven-membered D ring could be found.

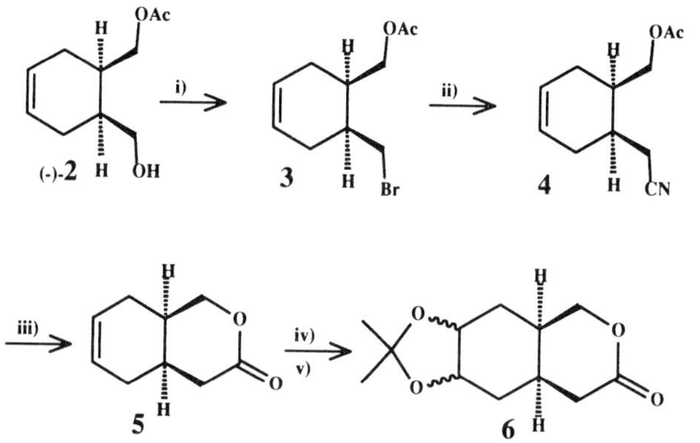

diastereomeric mixture of diols

i) CBr_4, PPh_3, CH_2Cl_2, 0° C; ii) KCN, DMSO, 60° C, 1.5h; 84%; iii) H_2O_2, NaOH, MeOH, 60° C, 2h; 86%; iv) $KMnO_4$, NaOH, tBuOH/H_2O; v) Acetone, PTSA cat., r.t., 30'; 74%.

SCHEME 4

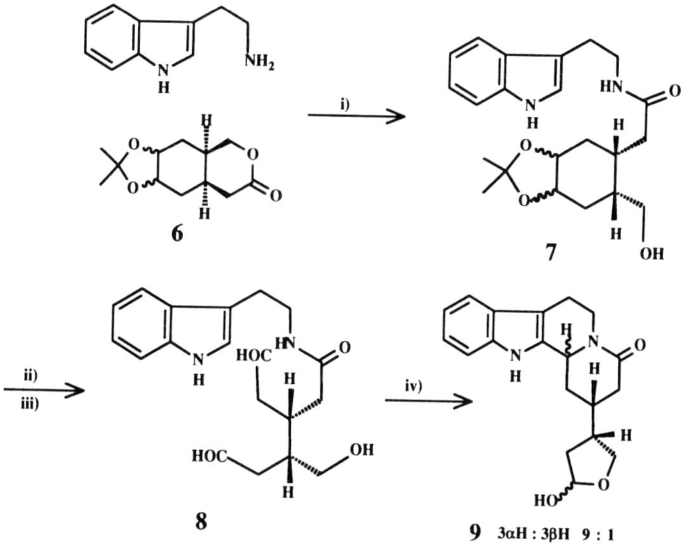

i) nBuOH, reflux, 8h; 81%; ii) HCl, MeOH-H_2O, r.t.; iii) $NaIO_4$, THF-H_2O, 0° C; iv) HCl 0.05 N, Acetone, 45° C, 15'.

SCHEME 5

The formation of **9** takes place through a favoured 6-exo-trig cyclization *via* an acyliminium intermediate. The dominant formation of the natural 3αH epimer may reflect a kinetically controlled intramolecular Pictet-Spengler reaction of the indolyl group to the iminium base from the less hindered β-face. Molecular models show that steric interaction between the axially oriented tetrahydrofuran appendage and the indole NH might disfavour the formation of the 3βH isomer.

The two lactams **9** showed very similar chromatographic properties and could only be separated with great difficulty on analytical scale.

Sequential reduction with $NaBH_4$ and $LiAlH_4$ (Scheme 6) produced the tetracyclic diol **10**, and at this stage the 3αH and 3βH epimers could be easily separated by column chromatography. Detailed spectroscopic investigations suggested that the two diols possess the conformation depicted in Figure 3. In particular, the major 3αH epimer shows no Wenkert-Bohlmann bands in the IR spectrum and a C-3H resonance at 3.91 ppm as a triplet (J = 5.0 Hz). For the minor 3βH epimer, *trans*-quinolizidine bands are present at 2840, 2800 and 2780 cm^{-1}, with a C-3H resonance upfield at 3.11 ppm as a doublet of doublets (J=11.0 and 1.2 Hz).

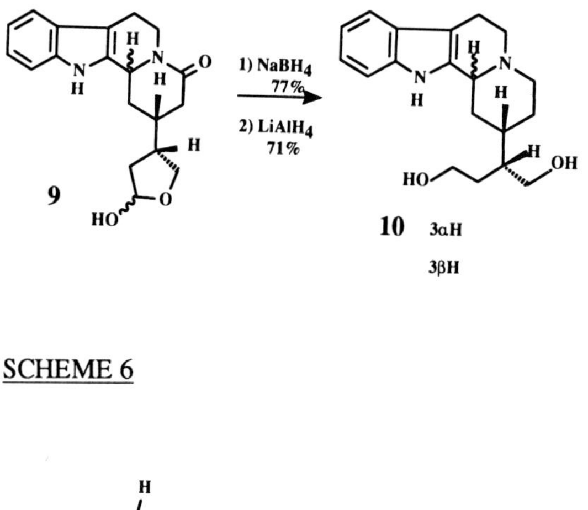

SCHEME 6

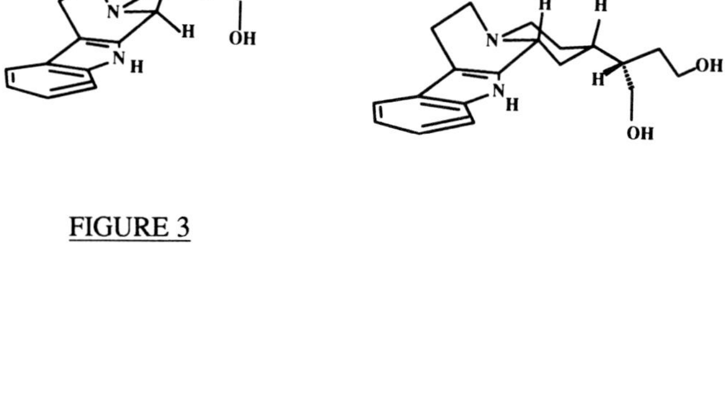

FIGURE 3

With the **10** (3αH) diol in hand, we envisioned the introduction of the double bond in the hydroxyethyl chain exploiting the Grieco protocol [7] (Scheme 7), which is based on the formation of the o-nitrophenylselenide **11**, periodate oxidation to the selenoxide **12**, followed by base-induced elimination.

With this sequence, (-)-antirhine, m.p. 111-113°C (Chf), $[\alpha]_D$= -1.96° (c=0.235 Chf), lit. [6]: m.p. 112-114°C (Chf), $[\alpha]_D$= -2.00° (c=0.23 Chf), identical in all spectroscopic properties with the natural sample, was obtained, although in very low yield. The main reason for the low yield was the lack of regioselectivity in the formation of the required phenylselenide **11**, which could only be isolated in 26% yield. In addition, internal nucleophilic displacement by the hydroxymethyl group during the subsequent elimination step, caused the concomitant formation of the tetrahydrofuran **13**, contributing to the low yield.

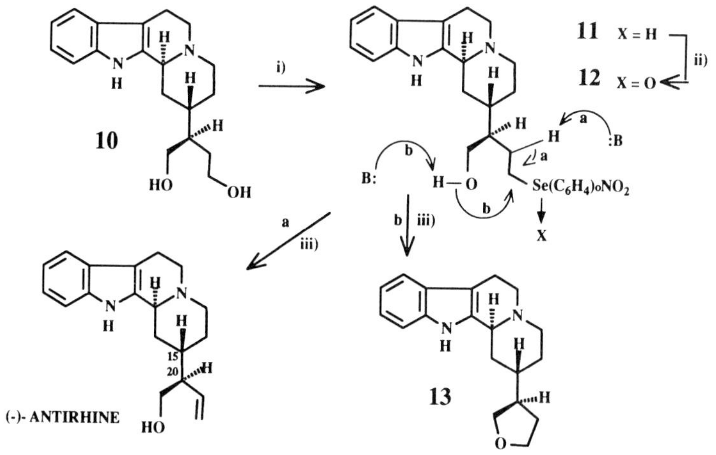

i) o-NO_2(C_6H_4)SeCN, Bu_3P, THF/DMF, r.t.; 26%; ii) $NaIO_4$ MeOH/ H_2O, 0° C; iii) Δ, Hünig's base : 40% + 36%.

SCHEME 7

These unsatisfactory results forced us to change the methodology for the introduction of the double bond in the molecule, and the Bamford-Stevens reaction [8] was therefore selected (Scheme 8).

Lactams **9** (3αH + 3βH) possess a masked aldehyde function which on reaction with p-TsNHNH$_2$ nicely furnished the p-tosylhydrazones **14** (R = H) as a mixture of diastereoisomers. Base treatment of **14** with NaH exclusively gave **15**

(17-oxo analogue of **13**), whereas the use of the more nucleophilic n-BuLi afforded a mixture of **15** and the alkylated adduct **16**. As a consequence, the hydroxymethyl function was protected to give **14** (R=TBDMS) which then underwent a smooth NaH-induced elimination to afford the lactams **17** (mixture of 3αH and 3βH) in very good yield. After reduction of amide carbonyl and deprotection, (-)-antirhine could be separated from its 3βH epimer 3-epi-antirhine which also is a natural product.

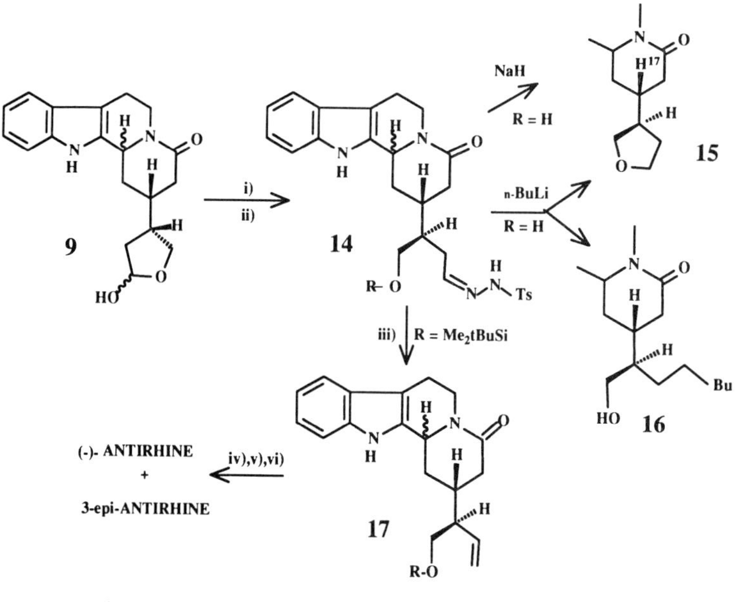

i) p-TsNHNH$_2$, MeOH/AcOH, r.t.;   ii) TBDMSCl, (iPr)$_2$NEt, DMF; 86%   iii) NaH, toluene, reflux; 70%;
iv) LiAlH$_4$, THF, 45° C;   v) H$_3$PO$_4$ 3%, r.t., 30';   vi) FC, SiO$_2$; 72%.

SCHEME 8

An approach to racemic antirhine starting from norcamphor has previously been described by Takano [9]. Subsequently, (-)-antirhine has been obtained by Takano and by Kametani starting from D-mannitol [10] and from (R)-1,2-isopropyliden-glyceraldehyde [11] resp. Our approach competes well with the published ones because it is expeditious, utilizes accessible starting materials and gives a satisfying overall yield (35% starting from lactam **9**, plus 4% of 3-epi-antirhine).

Akagerine, isolated from *Strychnos usambariensis* [12], is another unique tetracyclic indole alkaloid possessing an unusual perhydroazepine ring coupled to the tetrahydro-β-carboline moiety by a N_1-C_{17} aminal bond.

The recognition that C-17 and C-18 can be derived (Scheme 9) from the sp^2 carbons of a cyclohexene, leads to the pivotal retron which is the enantiomer of that previously used for the synthesis of antirhine. This lactone **20** = *ent*-**5** is accessible in a straightforward manner from the cyclohexene dimethanol monoacetate (+)-(1*S*,2*R*)-**2**. Therefore, the absolute configuration at C-15 is the one embodied in C-1 of the starting material, whereas those at C-3 and C-17 are under control of a suitable reaction sequence. Another synthetic challenge is the obtainment of the correct E-configuration for the trisubstituted double bond.

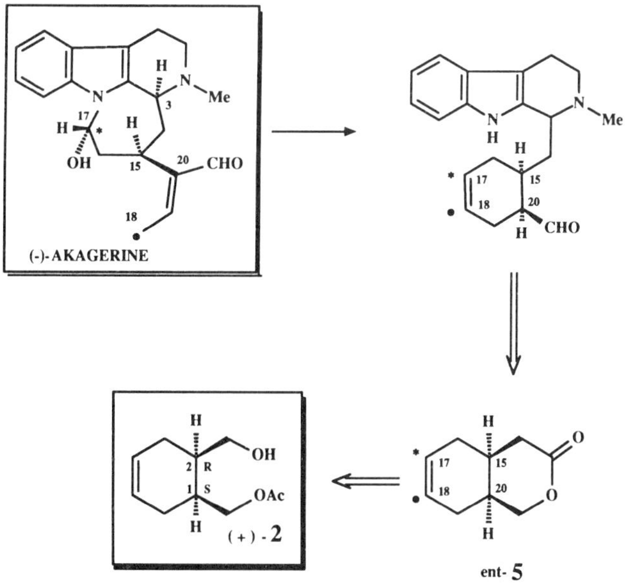

SCHEME 9

Asymmetrization of diacetate **1** with the enzyme PLE, as reported [4], gave variable results. Both chemical and optical yields of (+)-**2** were strongly dependent upon the preparation of enzyme. We have devised a better approach to (+)-**2**, applying the finding that PPL preparations were shown to have the same stereochemical preference (enantioselectivity) in aqueous medium for hydrolytic reactions, as for the "inverse" acylation reaction in anhydrous organic solvent [13]. As a consequence, from a meso or prochiral diol, using the same enzyme, a chiral monoacetate is produced by acylation of the diol in an organic solvent in the presence of a suitable activated ester, whereas the enantiomeric monoacetate is accessible by hydrolysis of the diacetate. Accordingly, by adding solid PPL to a solution of the diol **18** in anhydrous AcOEt (Scheme 10), a very high e.e. (>99.5%) was obtained when the reaction was stopped at 80% conversion to the monoacetate.

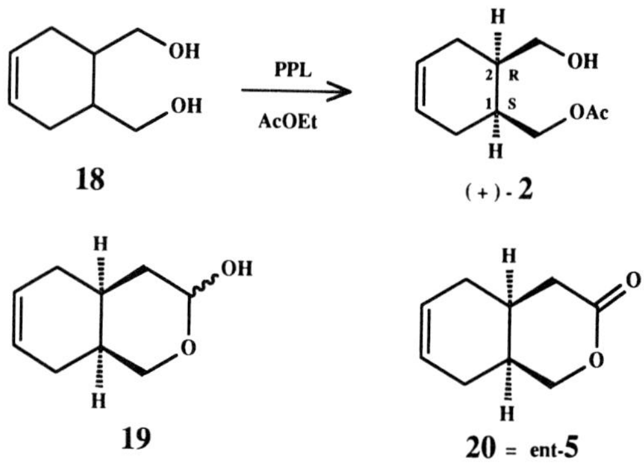

SCHEME 10

The lactone **20** (*ent-5*) was obtained from (+)-**2** *via* the lactol **19**. Compound **19** was reluctant to react with tryptamine even under forcing conditions, whereas **20** afforded the amide **21** (Scheme 11). Bischler-Napieralski cyclization of **21** gave the imine **22** which one should be able to reduce to a tetrahydrocarboline with the natural 3αH configuration. The initial results with less sterically demanding reducing agents such as NaBH$_4$ and NaBH$_3$CN in MeOH at r.t. were disappointing as reduction occurred without diastereoselectivity, affording the 3αH epimer **23** together with the "wrong" 3βH epimer **24** in a nearly 1 : 1 ratio. The two epimers were distinguishable by the sign of the Cotton effects in their CD curves: the 3αH epimer **23** showed a positive absorption and the 3βH epimer **24** a negative absorption at 267 nm, in agreement with literature data [14].

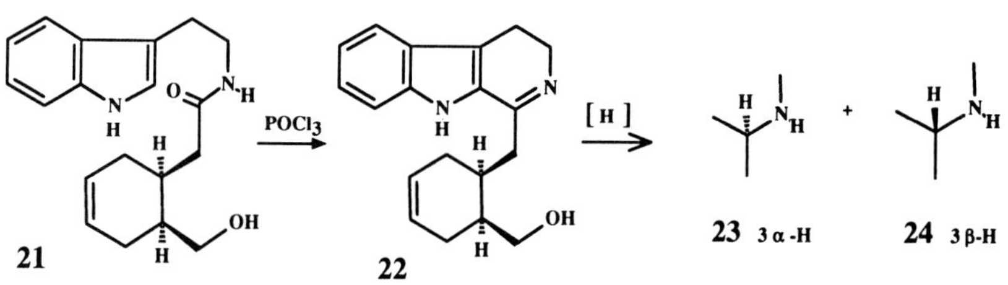

SCHEME 11

With the bulky DIBAH reagent in hexane or THF at -70°C, the

diastereoselectivity was shifted towards **24**, indicating that the β-face of the molecule is less hindered. The 1 : 1 ratio is restored in the presence of the Lewis acid $ZnCl_2$, probably because it forms a complex with the imine nitrogen which makes the less hindered face occupied before the reduction. A favourable result was obtained in the reduction with $Zn(BH_4)_2$ in THF at r.t. which afforded a fivefold excess of **23**, albeit with a low conversion.

Considering these facts and examining molecular models, we deduced that the desired 3αH diastereoselectivity could be improved by increasing the bulkiness of the cyclohexane appendage. To this end, the benzoate **25** (Scheme 12) was prepared and then reduced by $NaBH_4$ to afford a 13 : 7 mixture of **26A** to **27B**. A further increase to a 3:1 ratio was serendipitously obtained when 25.HCl in MeOH was added to a solution of $NaBH_4$ in MeOH at 0°C.

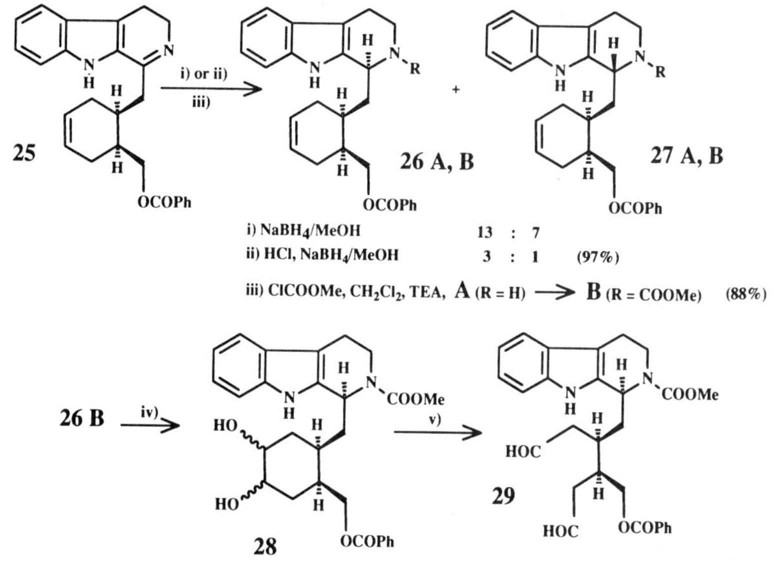
25
i) or ii)
iii)
26 A, B
27 A, B
OCOPh
i) NaBH4/MeOH 13 : 7
ii) HCl, NaBH4/MeOH 3 : 1 (97%)
iii) ClCOOMe, CH2Cl2, TEA, A (R = H) → B (R = COOMe) (88%)
26 B
iv)
v)
COOMe
HO
28
29
HOC

iv) OsO_4, THF/H_2O, NMO; 86%, v) $NaIO_4$, THF/H_2O, 0° C ; 96%.

SCHEME 12

Large scale separation between **26** and **27** could easily be performed on the *N*-carbomethoxyderivatives **26B** and **27B**, in which the urethane moiety serves as a protecting group for nitrogen in the subsequent oxidation of the double bond and as a precursor of the final *N*-methyl group.

The double bond of **26B** was then cleaved to the rather stable dialdehyde **29** (aldehyde protons resonating at 9.80 and 9.74 ppm, as br s) *via* the diol **28**. The dialdehyde was cyclized under mild conditions with dilute acid (Scheme 13) to the perhydroazepine **30**, which was identified by the disappearance of NH, the presence of one aldehyde signal and the appearance of an aminal proton at 6.26 ppm as a dd, J= 4.8 and 0.5 Hz. Product **31**, which is formed in variable amount by further intramolecular capture of the second aldehyde group (C18-H as a complex signal at 5.1 ppm), is always produced together with **30**.

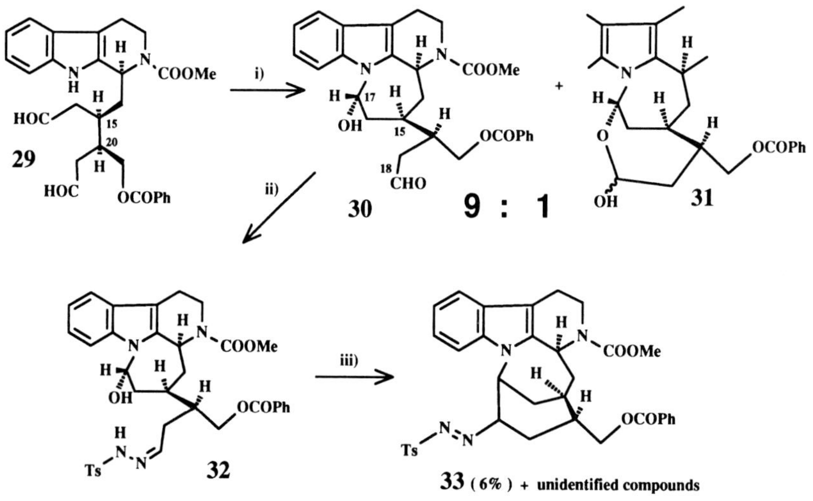

i) HCl 0.06 N, THF, 25°; ii) p-TsNHNH$_2$, THF/AcOEt, 30°; iii) NaH, Toluene, reflux.

SCHEME 13

It is worth noting that the cyclization is highly regioselective. Only products containing a seven-membered ring were found, and no eight-membered ring compounds derived from the alternative cyclization of the other aldehyde function were detected. Moreover, the stereogenic centre at C-17 is installed directly with the natural *S* configuration (as suggested by coupling constant values) with C17-H equatorial and C17-OH axial, as a consequence of steric interaction with aromatic C12-H. Therefore, all chiral centers of akagerine are introduced with the correct configuration and only the C(18)-C(19) double bond remains to be elaborated.

With the analogy of the synthesis of antirhine, we planned to prepare the tosylhydrazone **32** which should be able to undergo a NaH-induced elimination. To our surprise, this reaction afforded a complex mixture of products from which only the sulfonilazocompound **33** could be identified.

Therefore, we again turned to the Grieco protocol [7] for the introduction of the double bond (Scheme 14). Protection of C17-OH group as a methylether, reduction of the aldehyde to the primary alcohol followed by reaction with o-nitrophenylselenocyanate gave **34** and then **35**.

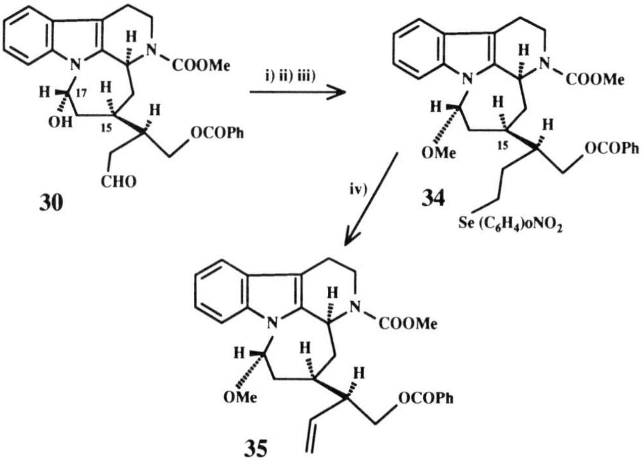

i) MeOH, HCl; ii) NaBH$_4$, MeOH; iii) o-NO$_2$(C$_6$H$_4$)SeCN, Bu$_3$P; iv) NaIO$_4$, MeOH/H$_2$O, 0° C; 93%.

SCHEME 14

The synthesis was concluded (Scheme 15) with the reduction of the urethane function to a methyl group, oxidation of the primary allylic OH to an aldehyde and finally deprotection of the C17 methoxy to a hydroxy group. Only the oxidation step requires some comments as it can be carried out with a variety of oxidizing agents. One of the best was found to be the Py-SO$_3$ complex in DMSO/TEA. Under these conditions, conjugation of the double bond occurs spontaneously to directly afford the natural E configuration. (-)-Akagerine, m.p.181-184°C, [α]$_D$= -16.30°, (c=1.06, MeOH), lit.: m.p. 188°C,[12] [α$_D$] = -16.6°,(c= 1, MeOH),[15] showed spectroscopic properties in complete agreement with the published data.

This alkaloid had previously been obtained in racemic form through an entirely different methodology[16].

The third compound synthesized from *cis*-cyclohex-4-ene-1,2-dimethanol diacetate is (+)-meroquinene (Scheme 16). This substance is a degradation product of some *Cinchona* alkaloids and is of great interest as it is the synthetic precursor of these therapeutically important bases[17].

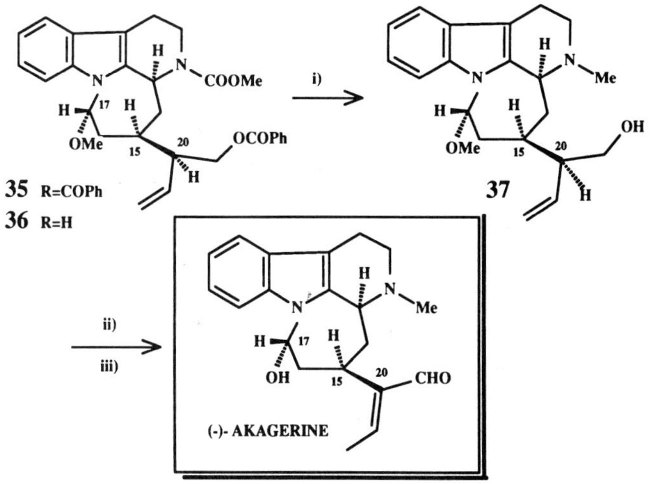

SCHEME 15

i) LiAlH₄, THF, reflux; ii) Py·SO₃, DMSO, TEA; iii) H⁺/H₂O, 45%

The structure of the target reveals the presence of two vicinal, functionalized chains at position 3 and 4 in a *cis*-relationship and *R* and *S* chirality, respectively. It is very easy to recognize that carbons 8 and 10 can be derived from the sp^2 carbons of a cyclohexene which in turn can be obtained from the known dimethanol monoacetate (-)-**2** which embodies the required chirality to generate the stereogenic centers C-3 and C-4.

The *cis*-octahydroisoquinoline **38** is accessible in excellent yield from (-)-**2** through conventional homologation, reduction and concomitant cyclization. The synthetic problem is now restricted to the cleavage of the double bond to form the dialdehyde **39** or the diester **40**. It is also necessary to devise a method that will differentiate the two carbonyl groups. To this end, many efforts have been made in order to intramolecularly trap one of the carbonyls with the NH of deprotected **39** or **40**. The aldehyde **39** (R=H) proved to be unstable affording only decomposition or polymeric products, whereas the diester **40** (R=H) was recovered unchanged when heated under different conditions (neat, in solvents and in presence of bases) or gave the diacid on hydrolysis.

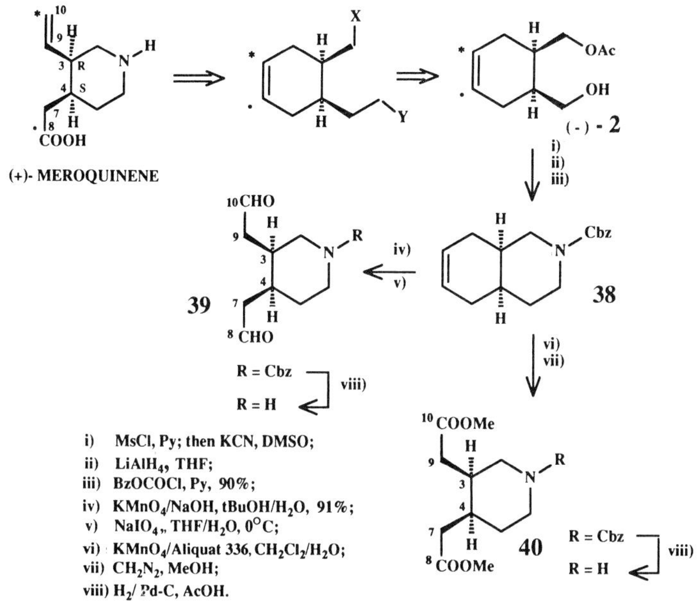

SCHEME 16

i)   MsCl, Py; then KCN, DMSO;
ii) LiAlH$_4$, THF;
iii) BzOCOCl, Py, 90%;
iv) KMnO$_4$/NaOH, tBuOH/H$_2$O, 91%;
v)   NaIO$_4$,, THF/H$_2$O, 0°C;
vi) KMnO$_4$/Aliquat 336, CH$_2$Cl$_2$/H$_2$O;
vii) CH$_2$N$_2$, MeOH;
viii) H$_2$/ Pd-C, AcOH.

In order to overcome these difficulties, we relied on the known ability of the enzyme PLE to hydrolyze esters[3] and we hoped that the specific interaction between enzyme and substrate would result in a discrimination of the chemically identical ester functions of **40**. Obviously, in this case, the enzyme behaves as a selective reagent and not as a chiral reagent.

We were very concerned about the results of this enzymatic hydrolysis (Scheme 17), which was performed by dissolving **40** in 12.5% DMSO/phosphate buffer at pH 7.25 and then adding solid PLE. After stirring for 20 hr, we reached a 95% conversion into a chromatographic single monoester which could be either **41** or the isomeric **44**. We were unable to differentiate **41** from **44** on the basis of the spectroscopic data. To check the structure, the compound was converted to a methyl vinylpiperidine acetate, which could have the "correct" **43** or the "wrong" **46** structure. After reduction of COOH to a primary OH followed by dehydration, the proton spectrum of the vinyl compound (Figure 4) was carefully analyzed and we found that the proton at 5.75 ppm is coupled to the methylene protons at 5.0-5.2 ppm (multiplet) with J = 17.0 and 11.0 Hz, and to a methine at 2.38 ppm (broad multiplet) through a large vicinal constant of 9.0 Hz. Moreover, the 2.38 ppm multiplet is coupled to two sharp and clean signals at 4.02 and 3.11 ppm, due to diastereotopic aminomethylene protons, with J = 4.0 and 2.8 Hz.

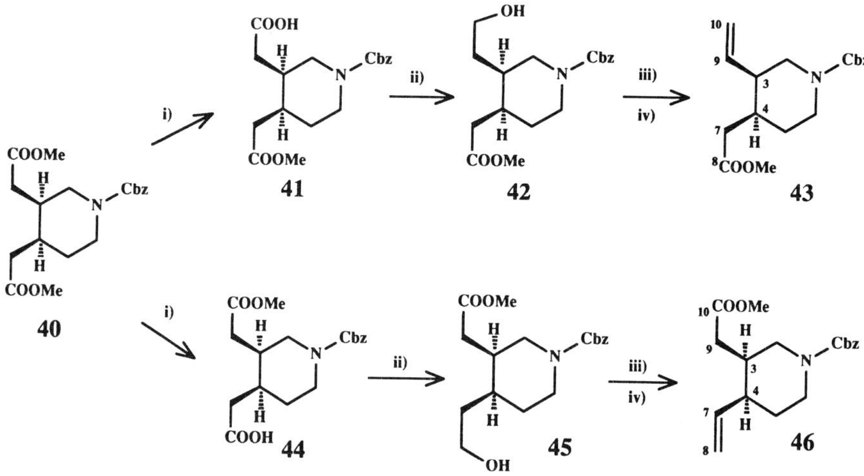

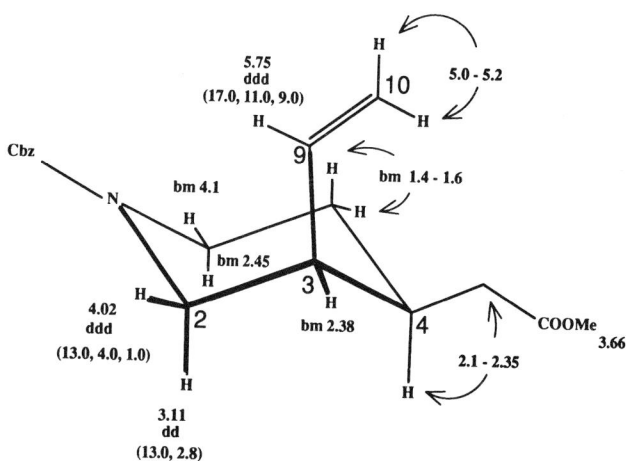

i) PLE/pH 7.25, phosphate buffer, 12.5% DMSO, 95%; ii) Me$_2$S-BH$_3$, THF, -78° C → 0°C, 60%
iii) o-NO$_2$(C$_6$H$_4$)SeCN, Bu$_3$P, THF, 74%; iv) NaIO$_4$, MeOH/H$_2$O, 70%

SCHEME 17

FIGURE 4

Chemo-enzymatic Approach to Some Indole Alkaloids 45

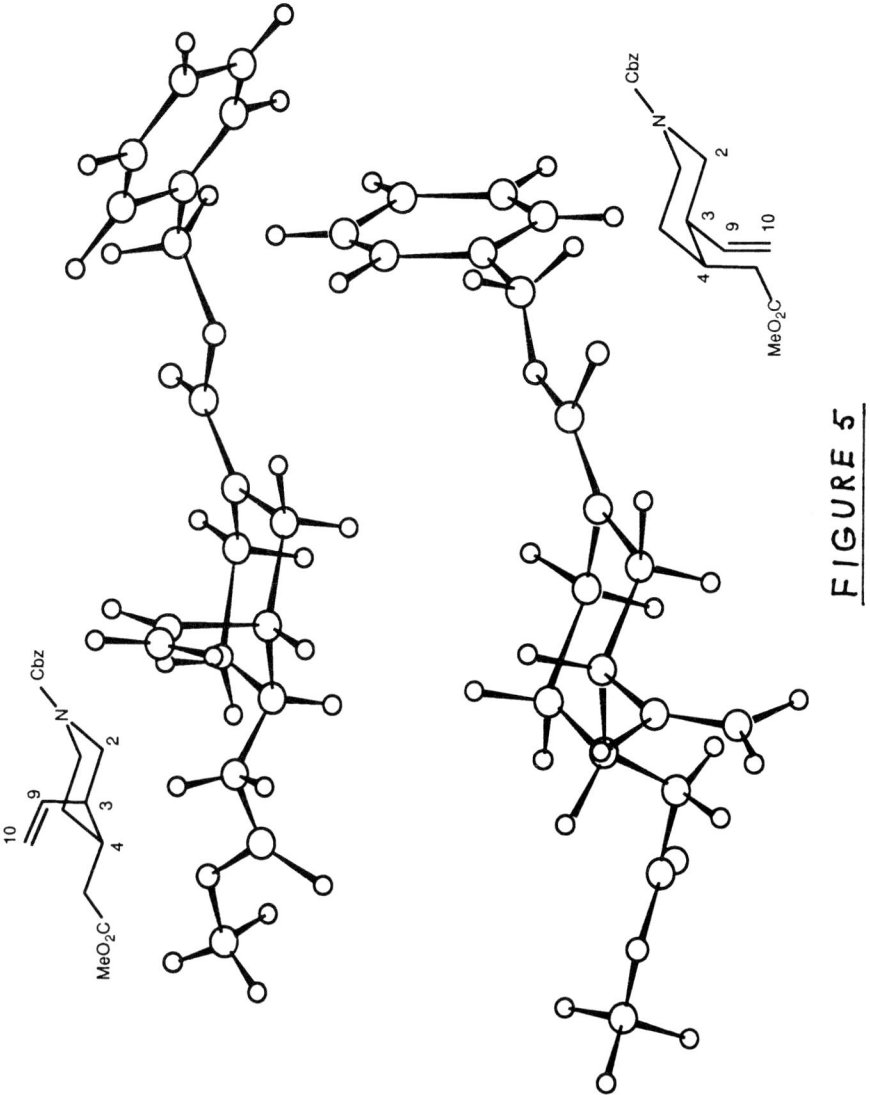

FIGURE 5

All these data indicate that the vinyl group is at position 3 of the piperidine ring in an axial orientation, and therefore the compound obtained has the desired structure **43**.

That the conformation of this meroquinene derivative is the one depicted in Figure 4 was confirmed by molecular mechanics calculations performed with the program DISCOVER (Biosym Technologies, Inc., San Diego). The calculation indicated (Figure 5) that in $CDCl_3$ (ϵ = 4.8), the conformer with an axial vinyl group and equatorial carboxymethylene moiety is 1.64 kcal/mol more stable than the corresponding conformer with an equatorial vinyl and an axial carboxymethylene. Other indications were that in the preferred conformation, the carbonyl of the Cbz-protecting group is opposite to the vinyl and that vinyl itself is almost eclipsed to the vicinal methine, thus explaining the large value of coupling constant.

Mild hydrolysis with 20% HCl then gave the target (+)-meroquinene isolated as a hydrochloride, which showed identical optical ($[\alpha]_D$ = +27.5°C, c = 0.6, MeOH) and spectral properties compared with those reported in the literature.

Some syntheses of meroquinene from achiral precursor have been reported[18] and in some of these cases resolution of an advanced intermediate led to both enantiomers of this acid[19]. A stereoconservative approach from secologanine has been reported too[20], and very recently a D-glucal derivative was chosen as starting material for an efficient synthesis[21].

3. SYNTHESIS OF ALKALOIDS FROM DIMETHYL *CIS*-3,5-BISMETHOXYCARBONYLPIPERIDINE. A PRELIMINARY APPROACH.

Alkaloids having the pseudo-*Aspidosperma* skeleton have been isolated quite recently from natural sources. Some representative examples, depicted in Scheme 18, are:
- 20*R*-15,20-dihydrocleavamine[22], which is structurally related to 16β-carbomethoxyvelbanamine, the indole portion of the antitumor bisindole alkaloids;
- 20*R*-1,2-dehydropseudoaspidospermidine[23], which is related to pseudotabersonine and pandoline;
- tacamonine[24], belonging to the pseudo-*Eburnea* class which is biogenetically and synthetically derived from the pseudo-*Aspidosperma* class through a well established oxidative rearrangement.

A rapid look at these structures immediately reveals that they possess the same *R* configuration at C-20. They share a common structural feature in that the disconnection of the non-indole portion of the molecules gives a *cis*-3,5-disubstituted piperidine as the ultimate synthetic precursor. More specifically, a suitable advanced precursor is *cis*-3-ethyl-5-piperidine acetic acid, which in turn could be derived from methyl hydrogen *cis*-3,5-piperidine dicarboxylate (C_1-symmetry) with the proper chirality at the stereogenic centers.

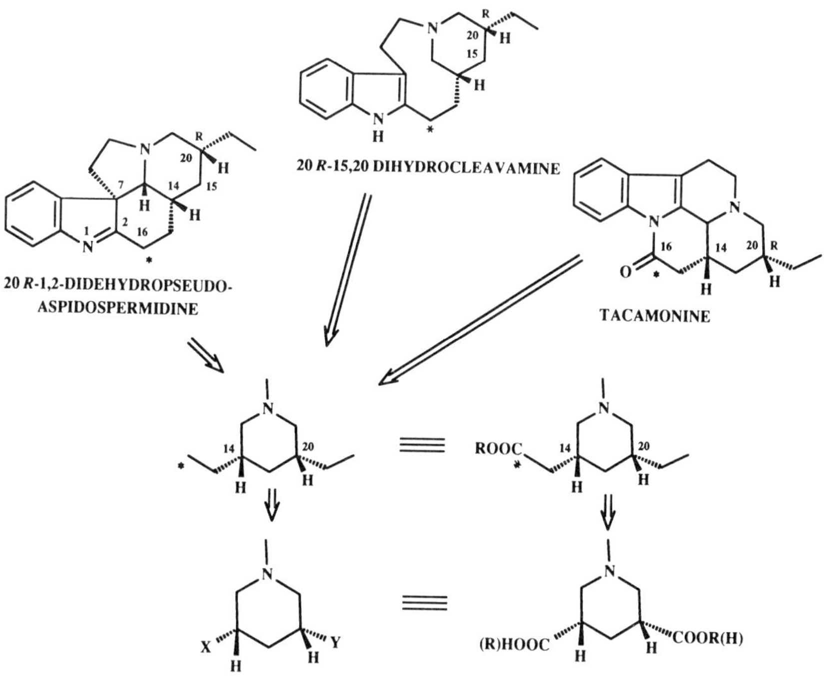

SCHEME 18

Therefore, the first step in the synthetic plan was to obtain the above mentioned compound in optically active form through enzymatic hydrolysis of the corresponding *cis*-3,5-bismethoxycarbonylpiperidine (σ-symmetry).

Enzymatic hydrolysis by means of PLE of a few similar compounds has appeared in the literature. The *R*-enantiomer is obtained in 90% yield and 40% e.e. from the dimethyl 3-aminoglutarate at pH 8 in 10% acetone solution/phosphate buffer if the substituent at nitrogen is H or a small group[25]. Similarly, the *S*-enantiomer is obtained if the substituent is large (R = Bz, 60% yield, 72% e.e.; R = Z, 93% yield and e.e.). Recently, an optically pure (e.e. 100%) (2*S*,5*R*)-*N*-benzyl-5-methoxycarbonylproline has been prepared, albeit in moderate chemical yield, *via* PLE catalysed hydrolysis of cis-N-benzyl-3,5-bismethoxycarbonylpyrrolidine in buffered aqueous 25% DMSO at pH 7.5[26].

Keeping this information in mind, we synthesized (Scheme 19) cis-N-benzyloxycarbonyl-3,5-bismethoxycarbonylpiperidine **48** by hydrogenation of the corresponding pyridine **47**, followed by separation of cis- and trans-isomers.

Enzymatic hydrolysis of **48** in 25% DMSO at pH 7.0 was stopped after 6 hr at 60% conversion to check the optical purity of the half-ester **49**.

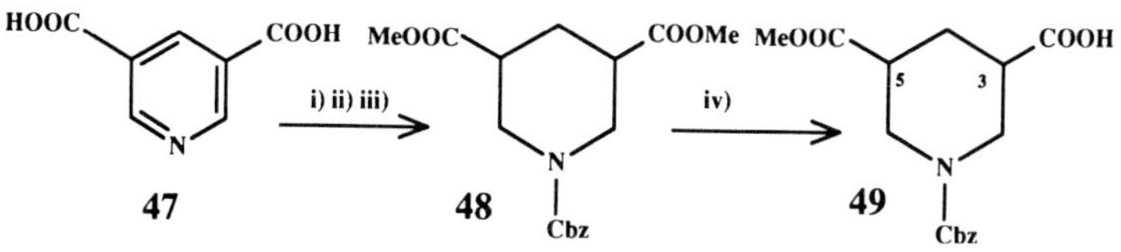

i) MeOH, PTSA, reflux; ii) H_2-PtO_2, AcOH; iii) CbzCl, Et_3N/Diox; FC, SiO_2;
iv) PLE/pH 7, phosphate buffer, 25% DMSO; 60%.

SCHEME 19

As usual ^{1}H-NMR methods were unsuccessful, the e.e. was determined by HPLC using an LKB Enantiopack column, eluting with a buffered solution at pH 6 (10mM NaH_2PO_4, 0.1M NaCl, 8% propan-2-ol). The e.e. was not very high (35-44%) but promising for future enzymatic studies and sufficient enough for our preliminary chemical studies.

Associated with the determination of e.e. is the problem of the determination of the absolute configuration of **49** which could be either 3R,5S or 3S,5R. Radical induced decarboxylation following the Barton protocol[27] would give a nipecotic acid derivative, and after hydrolysis an enriched (S)- or (R)-nipecotic acid[28]. However, the optical rotation of these acids is low and therefore we decided to assess the absolute stereochemistry of **49** by comparing the optical rotation of the final synthetic target (obtained through non-racemizing processes) with the one of the natural product of known chirality. Therefore, it was necessary to elaborate an enantio-divergent approach to some enantiomeric synthetic intermediates.

Initially, we planned to homologate the carboxy function to a carboxyme-thylene function (or equivalent) and to subsequently introduce the ethyl chain. These intermediates were identified as the nitrile esters **51** and ent-**51**. Nitrile **51** was prepared in excellent yield as depicted in Scheme 20. The carbomethoxy group was then elaborated to the phosphonate **53** in order to have access to the ethyl function via the corresponding vinyl. Unfortunately, all efforts to perform a Wittig olefination on **53** with formaldehyde in the presence of t-BuOK, resulted in formation of a complex mixture of products, from which we were unable to detect the desired vinyl compound.

On the other hand, when the carboxy function of **49** was protected as an orthoester (Scheme 20), the subsequent reduction of the methoxycarbonyl function with LiBH$_4$ gave **54** in only 26% yield together with substantial amount of products derived from hydrolysis of the ester function, the hydroxyacid itself and the *cis*-N-carbobenzyloxy-piperidine 3,5-dimethanol.

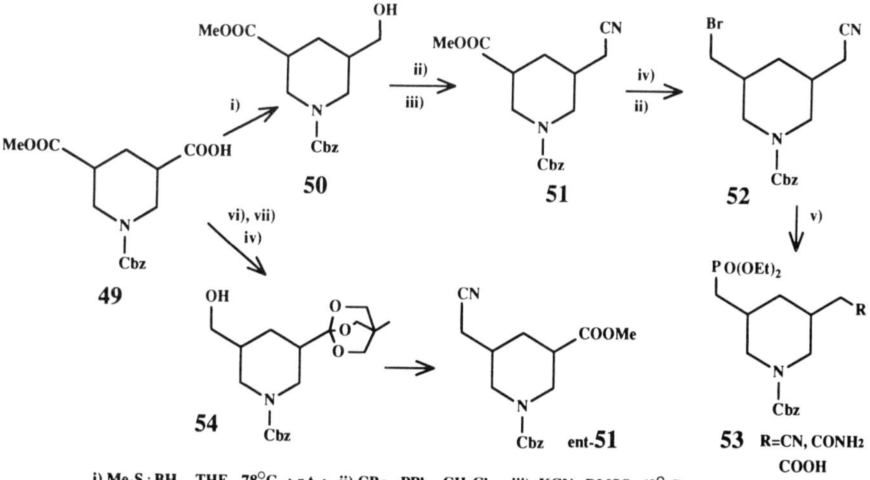

i) Me$_2$S · BH$_3$, THF, -78°C → r.t. ; ii) CBr$_4$, PPh$_3$, CH$_2$Cl$_2$; iii) KCN, DMSO, 60° C; iv) LiBH$_4$, THF; v) P(OEt)$_3$, 140° C; vi) (COCl)$_2$, DMF/CH$_3$CN, -20° C; 3-methyl-3-oxetanemethanol, Py; vii) BF$_3$ .Et$_2$O, CH$_2$Cl$_2$.

SCHEME 20

The above results indicated that some interaction had occurred between the two appendages, and this was further substantiated when we tried to introduce the ethyl chain by vinylation of the aldehyde obtained by oxidation of alcohol **50** (Scheme 21) by Wittig olefination. During the oxidation step we got a quantitative yield of an unseparable mixture of aldehyde, *cis*-**55** and *trans*-**55**, in a 48 : 52 ratio, as indicated by careful analysis of the NMR spectrum. In addition, the analysis indicated that in the *trans*-isomer, the aldehyde function is axially oriented.

That the *trans* isomer was thermodynamically more stable than the *cis* isomer was supported by molecular mechanics calculations (Discover). We found that *trans*-**55**, with axial aldehyde, equatorial methoxycarbonyl and the amide carbonyl directed towards the aldehyde, is by far the most stable among all possible stereoisomers and conformers. The conformer with equatorial aldehyde and axial methoxycarbonyl was much less stable (+2.8kcal/mol), in agreement with the NMR data. In *cis*-55, the diequatorial isomer is less stable than *trans*-55, but at least six conformations exist with energies ranging from +0.41 to +1.95 kcal/mol. *Cis*-55, with diaxial appendages, is by far the less stable structure. However, at least five conformations are present with energies between +1.65 and +2.73 kcal/mol. Taking into account the energy values and neglecting the contribution of isomers with exceeding energy of 3 kcal/mol, we could calculate (Figure 6) that aldehyde ester **55** exists as a thermodynamic mixture of *trans*- and *cis*-isomers, approximately in a 52 : 48 ratio, as suggested by NMR. The possibility of the coexistence of a diaxial conformer together with the diequatorial conformer in *cis*-**55**, might explain the behaviour described during the attempted formation of **54**.

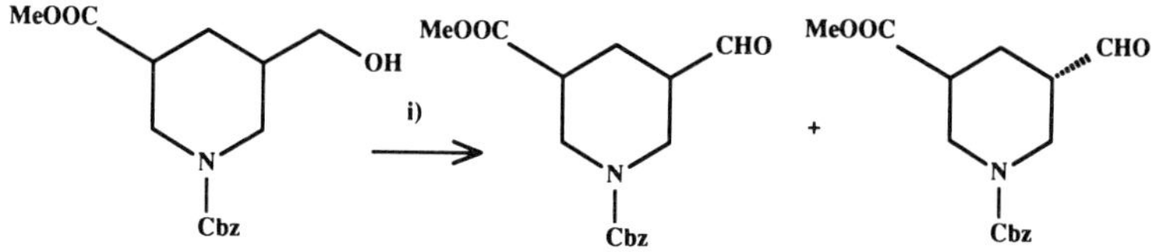

i) $(COCl)_2$, CH_2Cl_2, DMSO, Et_3N, -60° C.

SCHEME 21

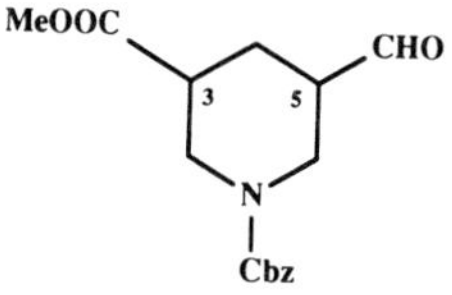

trans-(3 eq, 5 ax)	51.4
trans-(3 ax, 5 eq)	0.3
cis-(3 eq, 5 eq)	42.9
cis-(3 ax, 5 ax)	5.4

FIGURE 6

Continuing the program of introducing first the ethyl chain and exploiting the existence of *cis*-diaxial isomer, the lactone **56** (Scheme 22) was smoothly prepared from **49** and then reduced to the lactol **57**. To obtain the enantiomer of **57**, **49** was transformed into the *N*-methyl-*N*-methoxyamide **58** which in turn directly gave the lactol *ent*-**57** by DIBAH reduction.

Lactol **57** was converted by Wittig olefination, followed by KCN homologation, hydrogenation and hydrolysis to *cis*-5-ethyl-3-piperidine acetonitrile **60**, which possesses the carbon framework of the terpenoid portion of the targets depicted in Scheme 18.

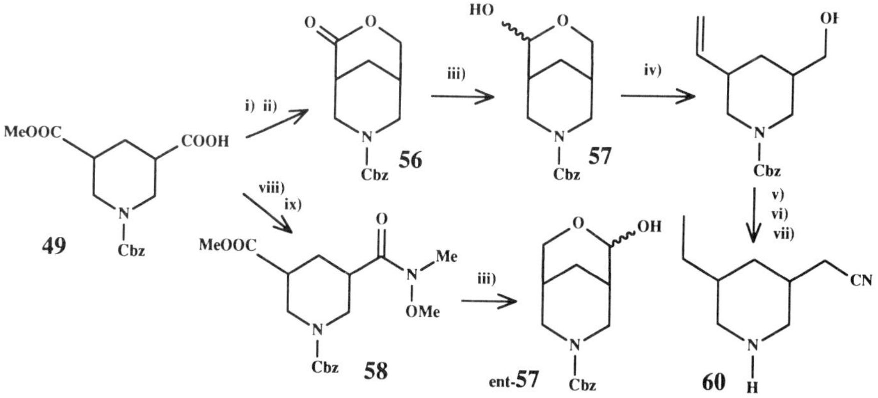

i) Me$_2$S-BH$_3$, THF, -78° → r.t.;
ii) PTSA, CH$_3$CN, 80°C;   iii) DIBAH, THF, -50°C;
iv) Ph$_3$PCH$_3^+$Br$^-$, DMSO, NaH   75%;   v) CBr$_4$, PPh$_3$, CH$_2$Cl$_2$ 0°C;
vi) KCN, DMSO, 60°C;   vii) H$_2$-Pd/C, AcOH,   56%;
viii) SOCl$_2$;   ix) N(OMe)(Me)H, CH$_2$Cl$_2$, 0°C, Py.

SCHEME 22

The reaction of **60** with tryptophyl bromide (Scheme 23) gave the *cis*-5-ethyl-*N*-tryptophyl-3-piperidine acetonitrile **61** which then afforded the nine-membered ketone **62** (M$^+$ 296 amu, UV λ_{max} 243 nm (log ϵ 4,25), 319 (4,3)) on cyclization. Reductive deoxygenation of **62** will lead to 15,20-dihydrocleavamine **63**.

Comparison of the optical rotation of this synthetic compound with that of natural 20R-15,20-dihydrocleavamine ([α]$_D$= + 133°, c = 0.78, Chf) will make it possible to determine the absolute configuration of **49**.

Further experimental work directed to improve the moderate optical yield and to obtain **63** from **62** is in progress.

Acknowledgements-This work was supported by CNR 'Piano Finalizzato Chimica Fine e Secondaria II'.

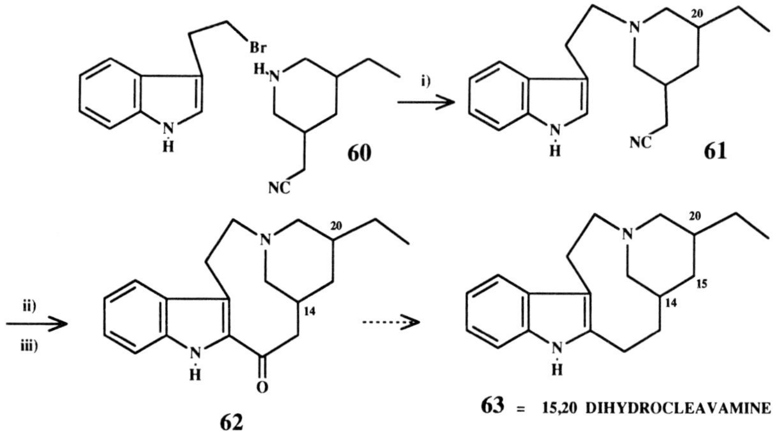

i) K_2CO_3, MeCN, 40°C; ii) NaOH 20%, reflux; iii) PPA, 80° C, 6h.

SCHEME 23

REFERENCES

1. M. Ohno, Enzymes in Organic Synthesis, Ciba Foundation Symposium, 1985, **111**, 171.
2. H.-J. Gais, K.L. Lukas, W.A. Ball, S. Braun and H.J. Lindner, Liebigs Ann.Chem., 1986, 687.
3. M. Ohno and M. Otsuka, Organic Reactions, Vol. 27, John Wiley & Sons, New York, 1 - 55.
4. R. Riva, L. Banfi, B. Danieli, G. Guanti, G. Lesma and G. Palmisano, J.Chem.Soc., Chem.Commun., 1987, 289.
5. K. Laumen and M. Schneider, Tetrahedron Lett., 1985, **26**, 2073.
6. R.S. Johns, J.A. Lamberton and J.L. Occolowitz, Aust.J.Chem., 1967, **20**, 1463.
7. P.A. Grieco, S. Gilman and M. Nishizawa, J.Org.Chem., 1976, **41**, 1485.
8. R.H. Shapiro and M.J. Heath, J.Am.Chem.Soc., 1967, **89**, 5734.
9. S. Takano, M. Takahashi and K. Ogasawara, J.Am.Chem.Soc., 1980, **102**, 4282.
10. S. Takano, M. Yonaga and K. Ogasawara, J.Chem.Soc., Chem.Commun.,

1981, 1155.
11. T. Suzuki, E. Sato, K. Unno, T. Kametani, Chem.Pharm.Bull., 1986, **34**, 3135.
12. L. Angenot, O. Dideberg and L. Dupont, Tetrahedron Lett., 1975, **16**, 1357.
13. G.M. Ramos Tombo, I.-P. Schär, X. Fernandez i Busquets and O. Ghisalba, Tetrahedron Lett., 1986, **27**, 5707.
14. W. Klyne, R.J. Swan, N.J. Dostoor, A.A. Gorman and H. Schmid, Helv.Chim.Acta, 1967, **50**, 115.
15. J. Udengene Oguakwa, C. Galeffi, M. Nicoletti. I. Messana, M. Patamia and G.B. Marini-Bettolo, Gazz.Chim.Ital., 1980, **110**, 97.
16. W. Benson and E. Winterfeld, Heterocycles, 1981, **15**, 935.
17. R. Veerporte, J. Schripsema and T. van der Leer, The Alkaloids (A. Brossi Ed.), Vol. 34, 331, Academic Press, New York, 1988.
18. a) M.R. Uskokowitz, J. Gutzwiller and T. Henderson, J.Am.Chem.Soc., 1970, 92, 203; b) S.Takano, M. Takahashi, S. Hatakeyama and K. Ozasawara, J.Chem.Soc.. Chem.Commun., 1979, 556.
19. M.R. Uskokowitz, T. Henderson, C. Reese, H.L. Lee, G. Grethe and J. Gutzwiller, J.Am.Chem.Soc., 1978, **100**, 571.
20. R.T. Brown and J. Leonard, J.Chem.Soc., Chem.Commun., 1978, 725.
21. S. Hanessian, A-M. Faucher and S. Leger, Tetrahedron, 1990, **46**, 231.
22. J.P. Kutney, J. Cook, K. Fuji, A.D. Treasurywala, J. Clardy, J. Fayos and H. Wright, Heterocycles, 1975, **3**, 205.
23. T.A. van Beek, R. Verpoorte, A. Baerheim Svendsen, Tetrahedron, 1984, **40**, 737.
24. T.A. van Beek, P.P. Lankhorst, R. Verpoorte and A. Baerheim Svendsen, Tetrahedron Lett., 1982, 4827.
25. M. Ohno, S. Kobayashi, T. Iimori, Y-F. Wang and T. Izawa, J.Am.Chem.Soc., 1981, **103**, 2405.
26. F. Björkling, J. Boutelje, M. Hjalmarsson, K. Hult and T. Norin, J.Chem.Soc., Chem.Commun., 1987, 1041.
27. D.H.R. Barton, Y. Herve', P. Potier and J. Thierry, J.Chem.Soc., Chem.Commun., 1984, 1298.
28. D. Pitre' and L. Fumagalli, Farmaco Ed. Sci., 1962, **17**, 130.

Sulfur-substituted Tryptophan — The Core of Phallotoxins, Virotoxins, and Amatoxins

H. Faulstich

BIOORGANIC RESEARCH GROUP, MAX-PLANCK-INSTITUTE FOR MEDICAL RESEARCH, HEIDELBERG, FEDERAL REPUBLIC OF GERMANY

1 INTRODUCTION

All toxic peptides produced by Amanita phalloides or A. virosa mushrooms contain a tryptophan residue substituted in α-position with a sulfur atom. In most cases the sulfur atom is part of a cysteine moiety resulting in an unusual amino acid, for which the name tryptathionine (Fig. 1) has been proposed[1]:

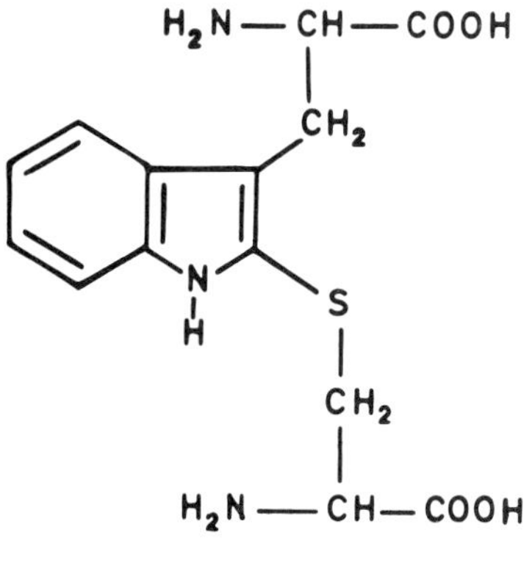

Fig. 1 Tryptathionine

In the phallotoxins cysteine and tryptophan are in juxtaposed positions of the cyclic heptapeptide. By bridging of the two side chains, a basket-like structure is formed with tryptathionine representing the handle (Fig. 2). It was shown that the bicyclic structure provides the rigid conformation of the phallotoxins and thus represents one of the prerequisites of their biological activity.

In the second family of cyclic peptides, the amatoxins, the bicyclic structure established through a linkage of juxtaposed cysteine and tryptophan moieties likewise is an essential prerequisite for the correct positioning of several other side chains involved in the binding of the octapeptide to its target protein. In the amatoxins the tryptathionine moiety has undergone further modifications, namely oxidation

of the sulfur atom and hydroxylation of the indole in
the 6'position (Fig. 3).

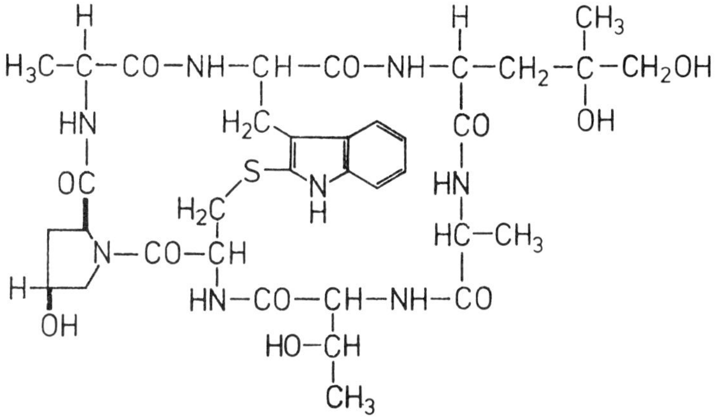

Fig. 2   Structure of phalloidin

The tryptathionine structures of phallotoxins and
amatoxins raised problems in structure elucidation, as
well as in the synthesis of the cyclopeptides from amino
acid building blocks by requiring chemical reactions not
hitherto known in traditional peptide synthesis. The
present study will summarize the unusual analytical and
synthetic properties of tryptathionine peptides. The
survey will cover not only phallotoxins and amatoxins but
also the virotoxins, a group of peptides from Amanita
virosa, which, despite a monocyclic structure, exhibit
biological activities very similar to those of the
bicyclic phallotoxins.

2 ELUCIDATION OF STRUCTURES

The structure of the main component of the phallotoxins, phalloidin, was elucidated by Th. Wieland and
coworkers[2,3]. By treatment with Raney-Ni the sulfur
atom was removed from the tryptathionine moiety, forming
a monocyclic peptide containing tryptophan and alanine
(dethiophalloidin). Incubation with mild acid (e.g. 50 %
trifluoroacetic acid) caused hydrolysis of the peptide
bond between the hydroxylated leucine and the alanine
residues, a reaction favored by the formation of a γ-
lactone ring. The linear peptide was then submitted to
Edman degradation. By rigorous treatment with Raney-Ni
it was possible to remove also the oxidized sulfur atom
of the amatoxins while the 6'hydroxygroup was simultaneously eliminated. In this case an octahydrotryptophan moiety

was formed allowing cleavage of the peptide ring at the hydroxylated isoleucine moiety and Edman degradation[4]. The two examples show that tryptathionine, either as thioether or sulfoxide, is easily cleaved by reduction in the presence of metal catalysts. In both cases the monocyclic peptides formed were devoid of biological activity.

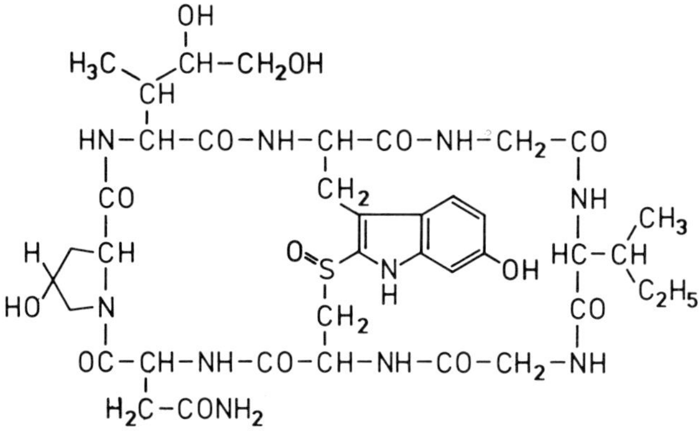

Fig. 3   Structure of α-amanitin

3 SYNTHESIS OF PHALLOTOXINS

By chemical modification of phalloidin at the dihydroxyleucine side chain (periodate cleavage followed by reduction of the keto group), the semisynthetic norvaline analog of phalloidin, norphalloin, was obtained with biological activity comparable to that of the mother compound. Norphalloin was the subject of the first synthetic approach in the field of phallotoxins[5]. The key reaction in this synthesis was the formation of the tryptathionine core, which was achieved by activating the thiol group of a cysteine-containing peptide as sulfenyl chloride by reaction with N-chlorosuccinimide in acetic acid, and reacting it with the tryptophan moiety of a peptide containing the remaining amino acids. The H-shaped intermediate (Fig. 4) could be isolated in relatively good yield. It is important to note that activation of thiols by N-chlorosuccinimide must be performed in the absence of tryptophan, because otherwise chlorination in ß-position of the indole part will be the prevailing reaction[6]. Deblocking and cyclization of the intermediate compound finally yielded norphalloin which was identical to the naturally derived product with respect to physical, chemical and biological properties. In the following, this successful synthetic route was used for the preparation of several norphalloin analogs which allowed information

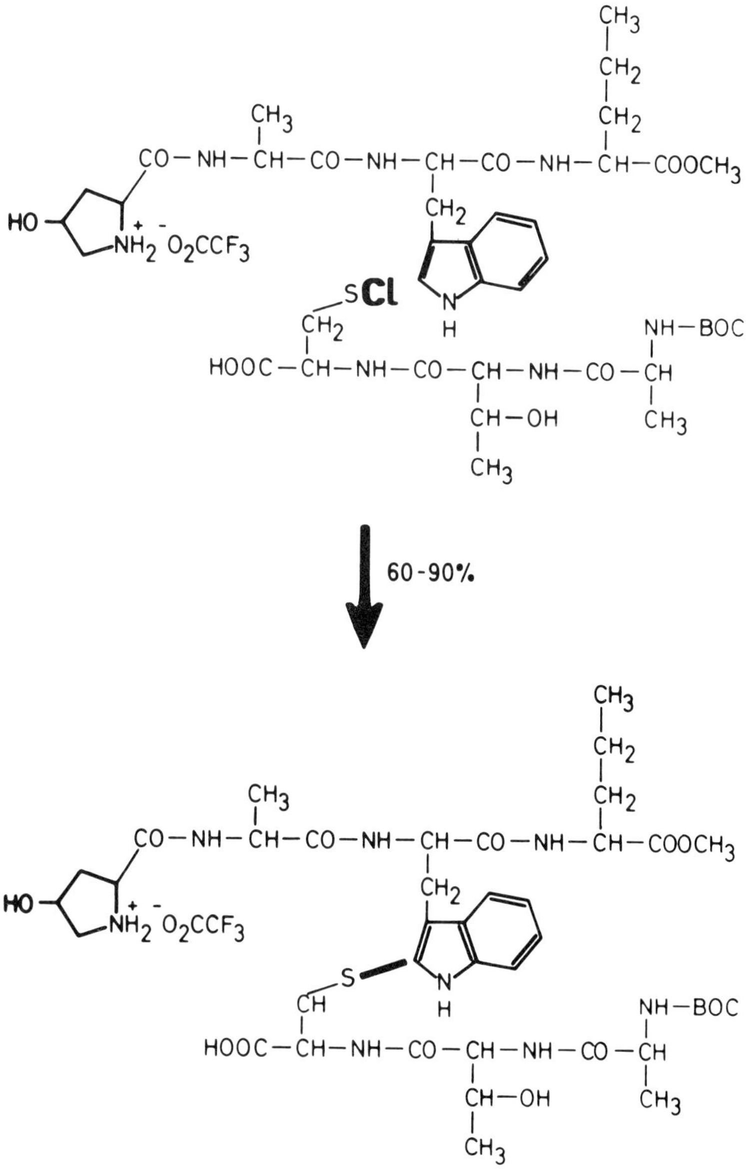

Fig. 4   Synthesis of norphalloin: Formation of the
         tryptathionine moiety by the sulfenylchloride
         method

to be collected on the relationship between structure of
the phallotoxins and their biological activity. The
toxins bind to the cytoskeletal protein actin in its
polymeric form; by stabilization of the filaments in

cells they shift the equilibrium between monomers and polymers in favor of the polymers[7]. We were able to show that, for formation of this complex with actin, phallotoxins require all structural elements of the small peptide ring, such as the methylgroup of alanine, the allo-positioned hydroxygroup of proline, and, very probably, the tryptathionine group held in a correct position by the bicyclic structure[8, 9]. In contrast, the larger peptide ring is of minor importance for the biological activity of phallotoxins; particularly the hydroxylated leucine moiety appears to be located opposite to the actin binding site.

4 SYNTHESIS OF AMATOXINS

Amatoxins bind to a transcription enzyme, DNA-dependent RNA polymerase II (or B). Through this complexformation, the elongation step of transcription is blocked, leading to an inhibition of protein synthesis[10]. In contrast to the phallotoxins, the biological activity of amatoxins strictly depends on the presence of the hydroxylated isoleucine residue, particularly on the trans-configurated, γ-positioned hydroxyl group. Replacement of this hydroxylated isoleucine with an amino acid having only a hydrocarbon chain, e.g. isoleucine, reduced biological activity to ca. 1 %. Nevertheless the isoleucine analogs of α-amanitin (i.e. derivatives of the natural product amanullin) proved to be excellent models for studying the chemical reaction leading to the formation of the tryptathionine moiety[11]. It was found that the activating oxidation reaction need not necessarily be carried out with the thiol group of cysteine, but can also occur at the indole part of tryptophan. The basis of this reaction is the observation of Savige and Fontana[12] that tryptophan in the presence of peracetic acid forms an epoxide that rearranges to the hexahydropyrrolo indole derivative (Fig.5)

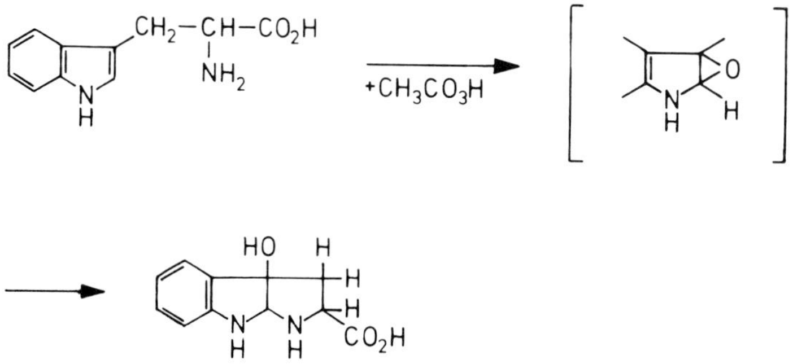

Fig. 5   Reaction of tryptophan with peracetic acid

In acidic medium the hexahydropyrrol ring is cleaved leaving the indole oxidized and activated for the addition of thiols. Since thiol groups can be protected by acid-labile protecting groups both steps, the activation of the indole and the deblocking of SH, can occur in one reaction. This allows the formation of tryptathionine in a single step, starting from one peptide chain containing both amino acids, tryptophan and cysteine (Fig. 6).

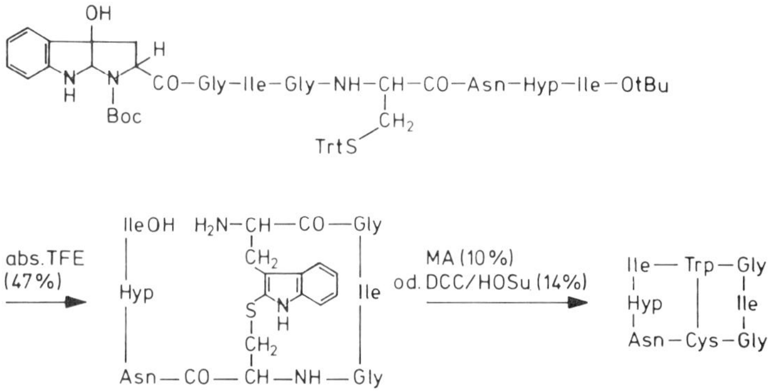

Fig. 6 Synthesis of isoleucine analogs of amanitin according to Zanotti et al.[11]

Using this reaction Zanotti and Wieland[11] prepared a great number of amanitin analogs. Unfortunately it proved to be very difficult to prepare hydroxylated isoleucintryptophan peptides with the γ-hydroxy group of isoleucine in the correct (R) configuration. All isoleucine analogs synthesized as model compounds of α-amanitin had inhibition capacities for RNA polymerase II that were lower by two orders of magnitude.

5 VIROTOXINS

About 10 years ago we found that the pattern of toxic peptides present in the white species Amanita virosa differs from that of A. phalloides. The white toadstool is as poisonous as the green one but contains only α-amanitin instead of the whole family of amatoxins[13]. (Unlike A. virosa mushrooms collected in Europe, specimens found in Virginia (USA) contained amaninamide, which is α-amanitin lacking the 6'hydroxygroup at the tryptathionine moiety[14]).

Differences between A. phalloides and A. virosa were also seen in the phallotoxins, where all acidic components (e.g. phallacidin), which in A. phalloides make up more than 50 % of all the phallotoxins, were

lacking. Instead we found a family of new toxins, the
virotoxins[15]. As seen from a direct comparison of the
main component of this family, viroidin, with the main
component of phallotoxins, phalloidin, ca. 50-60 % of
the structures are identical. The main difference between
the two toxins is the absence of tryptathionine in
viroidin. Instead, viroidin has a methylsulfonyl group
in the α-position of tryptophan and a D-serine residue
instead of L-cysteine (Fig. 7).

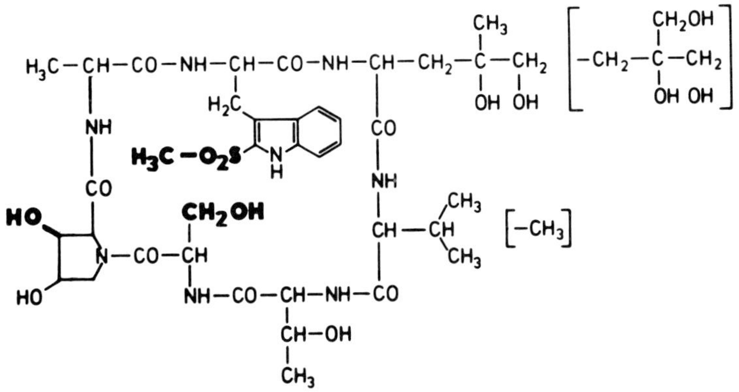

Fig. 7   Structure of virotoxins

It has been speculated that viroidin may represent
a metabolic product of an unknown tryptathionine-con-
taining precursor. Indeed it is tempting to suppose that
methylation of the sulfur atom in tryptathionine followed
by β-elimination of the sulfonium would be able to cleave
the tryptathionine bridge with formation of dehydroalanine
and the α-methylthioether of tryptophan. Oxidation of this
thioether and addition of water to dehydroalanine would
create the structure found in viroidin. Although monocyclic in nature, viroidin binds to actin with an affinity
comparable to phalloidin. It was shown that at least one
of the two additional hydroxy groups - one in 3,4 dihy-
droxyproline[16] and one in D-serine - is essential for
binding. It appears to us that in viroidin the disad-
vantage of a less rigid conformation is compensated for
by the presence of these two hydroxy groups, which
possibly contribute to actin binding by the formation of
an additional hydrogen bond.

6 IDENTIFICATION OF THE TOXINS

The tryptathionine moiety as in the phallotoxins, as well
as its derivatives in the other toxin families, are of
great importance for the identification of the toxins on
thin layer or paper chromatograms. The α-thioether of
tryptophan reacts with cinnamic aldehyde in the presence

of HCl vapor by developing a bright blue colour, thus
providing the usual method for the identification of
phallotoxins. Oxidation of the sulfur atom to a sulfoxide,
as for example in amaninamide, retards the reaction and
changes the colour to greyish brown. Sulfone-substituted
tryptophan derivatives like viroidin give no colour reaction at all.

Hydroxylation in the 6'position turns the colour to
purple and enhances the sensitivity by ca. 10 fold,
independent of the oxidation state of the sulfur atom.
This allows the determination of amatoxins in amounts as
small as 0.1 µg per spot, for example, on silica thin
layers. Amatoxins represent the most dangerous toxic
peptides in Amanita mushrooms and are entirely responsible for fatal cases of human poisoning. Therefore it
was desirable to have a simple procedure for detecting
the poisons in mushrooms extracts. Wieland et al.[17]
described such an assay in which a drop of mushroom juice
is dried on normal newspaper (from wood pulp) and developed with a drop of concentrated hydrochloric acid. Under
these conditions amatoxins develop within a few minutes
a blue colour, most probably by reaction with a dialkoxy-
substituted cinnamic aldehyde which is one of the numerous
structure elements of lignin.

7 SPECTRAL PROPERTIES

Substitution with a sulfur atom as in tryptathionine
shifts the u.v. absorption bands of tryptophan ca. 10 nm
to longer wavelengths. The typical shoulder around 300 nM
allows the spectral detection of phallotoxins even in the
presence of proteins. Oxidation of the sulfur as in sulfoxides or sulfones shifts the maximum of absorption
back to shorter wavelengths and changes the shapes of
the spectra in a typical way as illustrated in Fig. 8.

Although the peptide environment of such chromophores
may cause minor modulations, the spectra representing
the various oxidation states of tryptathionine can be
easily distinguished in all natural and synthetic toxins.
A typical u.v. absorption spectrum is also exhibited by
the chromophore of the amatoxins (Fig. 7). Due to the
bathochromic shift caused by hydroxylation in the 6'
position it possesses a strong absorption band at 310 nm.
In general, the u.v. spectra of all these toxic peptides
proved to be of great practical value, not only in
allowing the identification of the various toxins in
solution (e.g. column eluates) but also in permitting
exact usage of the toxins on the basis of molar extinction
coefficients (Table 1).

Some of the toxins form anions in alkaline solutions
as concluded from the strong bathochromic shifts of their
u.v. spectra. For example, amatoxins in alkaline solution
have an absorption maximum around 330 nm. However, ama-

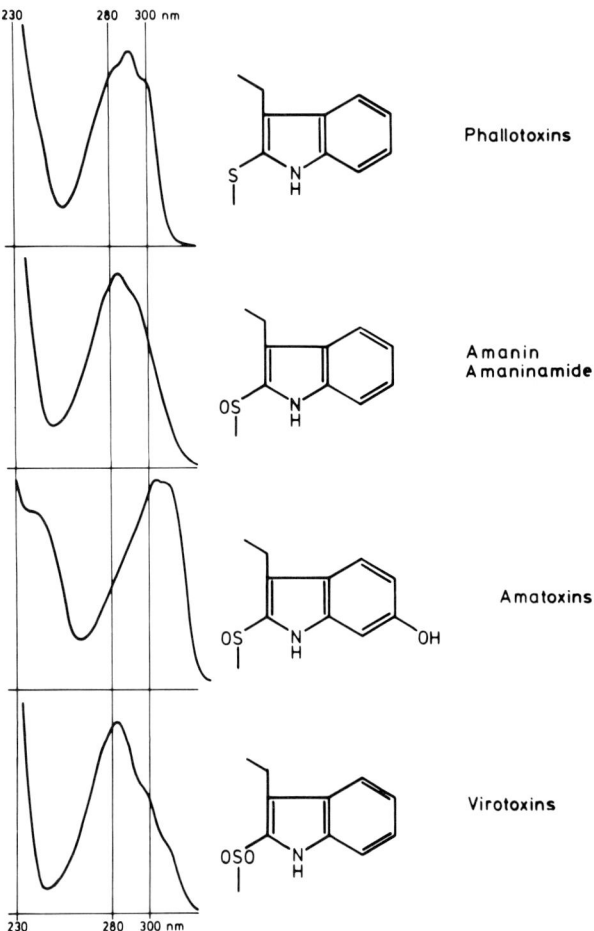

Fig. 8   U.V. absorption spectra of Amanita toxins

toxins are unstable in alkaline solutions, and the absorption band at 330 nm slowly disappears due to a breakdown of the chromophore by ß-elimination. Anion formation and shift of the u.v. spectrum was also observed with the virotoxins. Although these compounds are sulfones and hence even more susceptible to ß-elimination than the sulfoxides in the amatoxins, virotoxins withstand even treatment with strong alkali since they lack ß-positioned protons. Besides its typical (^{1}H) nmr signal, it was the absence of ß-elimination that revealed the presence of a methyl-sulfonylgroup in the virotoxins.

Table 1: U.V. absorption bands used for the identification or quantification of Amanita toxins (in water)

Phallotoxins	ε_{300}	$= 10.100 \, [M^{-1} cm^{-1}]$
Amatoxins pH 7	ε_{310}	$= 13.500$
pH 11	ε_{330}	$= 15.300$
Amaninamide	ε_{287}	$= 12.500$
Virotoxins pH 7	ε_{276}	$= 13.400$
pH 11	ε_{298}	$= 10.300$

8 CHIRALITY OF TRYPTATHIONINE CHROMOPHORES

Similar to cystine moieties when existing in a fixed position e.g. as part of a cyclic peptide or a protein, tryptathionine can possess helical chirality with Cotton effects in the range between 280 and 310 nm.

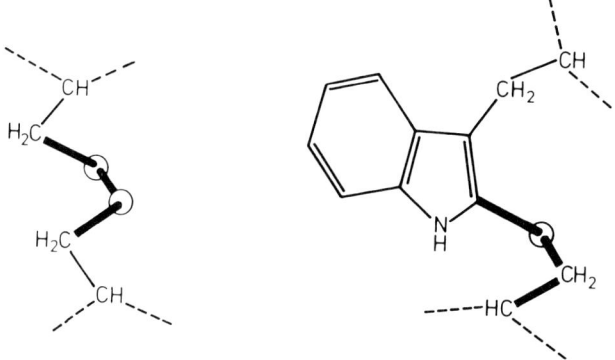

Fig. 9   Helical chirality of cystine and tryptathionine in cyclic peptides

According to the absolute configuration of the helix the Cotton effects can be positive or negative. In an early study of this kind[18], O.R.D. measurements had revealed that all phallotoxins with biological activity had positive Cotton effects between 280 and 310 nm and in the range of 250 nm. These findings were confirmed later on by C.D. measurements, for example of phalloin, a natural phallotoxin containing L-hydroxyleucine instead of L-dihydroxyleucine (Fig. 10).

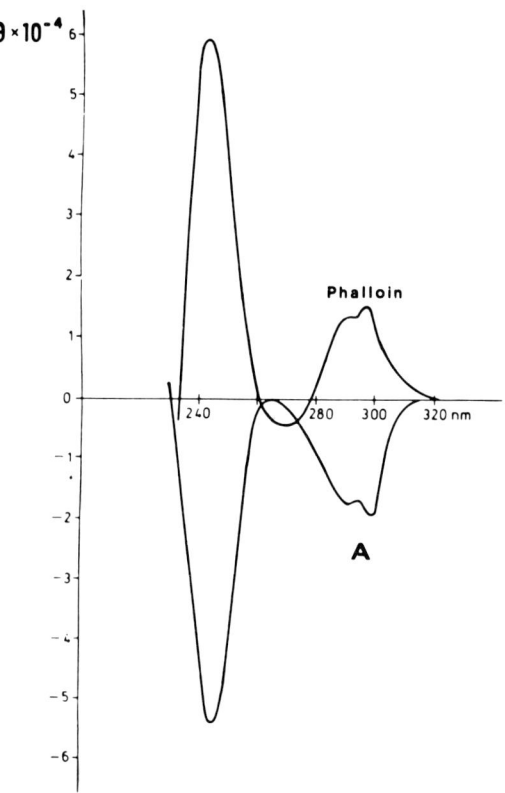

Fig. 10   C.D. spectra of a natural phallotoxin (phalloin) and a synthetic model compound (A) (see text)

It was shown that the positive Cotton effects of tryptathionine in the range of 280 to 310 nm are correlated to tryptathionine fixed in a P-helical configuration, defined as a clockwise screw[19]. A monocyclic, truncated phallotoxin model compound (A) synthesized by G. Zanotti (Fig. 11), which by X-ray analysis showed an M-configurated (anti-clockwise) helical arrangement of its tryptathionine part, exhibited a C.D. spectrum which was the mirror-image of that of phalloin.

Oxidation of the sulfur atom in tryptathionine, resulting in a mixture of sulfoxides in the (R) and (S) configuration, introduces another center of chirality. The oxidation products of phalloidin could be separated by chromatography and, as expected, (R) and (S) sulfoxide showed mirror-image C.D. spectra (Fig. 12). Only the (R) configuration, which exhibits positive Cotton effects around 300 nm, has high biological activity, while binding to actin and toxicity of the phalloidin sulfoxide with

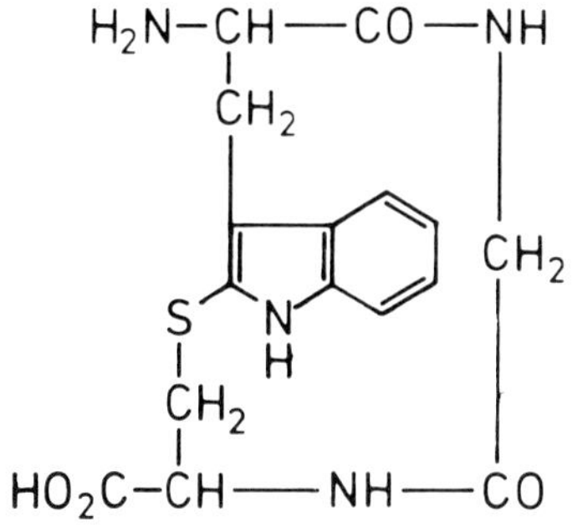

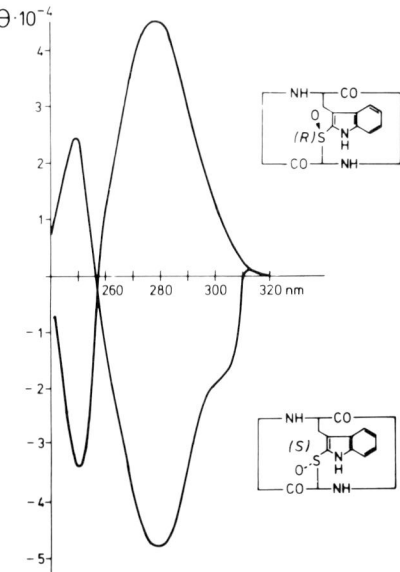

Fig. 11

Compound A, a tryptathionine model peptide exhibiting negative Cotton effects (M-chirality) according to ref. 19

the (S)-configuration were found to be reduced to ca. 10 %. Obviously, the strong dipole of the (S)-configurated sulfoxide disturbs the interaction with actin or induces conformational changes in the peptide backbone that are deleterious to biological activity. It is interesting to note that some minor components of the virotoxin family exhibiting biological activity but having a methylsulfinyl group instead of methylsulfonyl, likewise possess an (R) configurated sulfoxide moiety.

Fig. 12   C.D. spectra of the diastereometric sulfoxides obtained by oxidation of phalloidin

In the family of amatoxins also, the sulfoxide appears to be in the (R) configuration. This was concluded from the positive Cotton effects of α-amanitin in the range from 240 to 320 nm (no picture). A. Buku succeeded in reducing a 6'methylderivative of α-amanitin and reoxidizing it to a mixture of the sulfone and the (R) and (S) sulfoxides. Again, the (S) sulfoxide derivative showed poor biological activity while the full activity was restored to the (R) configurated compound. In contrast, the absence of a chiral center at the sulfur atom, as in the thioether compound or the sulfone of α-amanitin, did not significantly reduce the biological activity of the natural compound[20].

REFERENCES

1. Th. Wieland, K. Freter and E. Gross, Liebigs Ann. Chem. (1959) 626, 154

2. Th. Wieland and W. Schön, Liebigs Ann. Chem. (1955) 593, 157

3. Th. Wieland and H.W. Schnabel, Liebigs Ann. Chem. (1962), 657, 225

4. Th. Wieland and U. Gebert, Liebigs Ann. Chem. (1966) 700, 157

5. F. Fahrenholz, H. Faulstich and Th. Wieland, Liebigs Ann. Chem. (1971) 743, 83

6. Th. Wieland, Ch. Jochum and H. Faulstich, Liebigs Ann. Chem. (1966) 727, 138

7. H. Faulstich, A.J. Schäfer and M. Weckauf, Hoppe-Seyler's Z. Physiol. Chem. (1977) 358, 181

8. H. Faulstich, "Chemistry of Peptides and Proteins", (W. Voelter, E. Wünsch, J. Ovchinnikov, V. Ivanov eds.) Walter de Gruyter, Berlin (1982) Vol. 1, 279

9. Th. Wieland "Peptides of Poisonous Amanita Mushrooms", Springer, Heidelberg (1986)

10. A.C. Vaisius and Th. Wieland, Biochemistry (1982) 3097

11. G. Zanotti, C. Möhringer and Th. Wieland, Int. J. Peptide Protein Res. (1987) 30, 450

12. W.E. Savige and A. Fontana, Int. J. Peptide Protein Res. (1980) 15, 102

13. H. Faulstich, A. Buku, H. Bodenmüller, H. Dabrovski and Th. Wieland, "Structure and Activity of Natural Peptides" (W. Voelter and G. Weitzel, eds.) Walter

de Gruyter, Berlin (1981), 189

14. A. Buku, Th. Wieland, H. Bodenmüller and H. Faulstich Experientia (1980), 36, 33

15. H. Faulstich, H. Bodenmüller, A. Buku, J. Dabrovski and Th. Wieland, Biochemistry (1980) 19, 3334

16. A. Buku, H. Faulstich, Th. Wieland and J. Dabrovski, Proc. Natl. Acad. Sci., USA (1980) 77, 2370

17. Th. Wieland, L. Wirth and E. Fischer, Liebigs Ann. Chem. (1949) 564, 152

18. H. Faulstich and Th. Wieland, Europ. J. Biochem. (1971) 22, 79

19. Th. Wieland, B. Beijer, A. Seeliger, J. Dabrovski, G. Zanotti, A.E. Tonelli, A. Gieren, B. Dederer, V. Lamm, E. Hädicke, Liebigs Ann. Chem. (1981), 2318

20. A. Buku, R. Altmann and Th. Wieland, Liebigs Ann. Chem. (1974), 1580.

Chemistry and Biology of Ascorbigens

M. N. Preobrazhenskaya, A. M. Korolev, I. L. Plikhtyak,*
I. V. Yartseva,* S. A. Efimov,* E. I. Lazhko, and
L. G. Aleksandrova

INSTITUTE OF NEW ANTIBIOTICS OF THE USSR ACADEMY OF MEDICAL SCIENCES, 119867, MOSCOW, B. PIROGOVSKAYA 11, USSR * ALL-UNION CANCER RESEARCH CENTRE OF THE USSR ACADEMY OF MEDICAL SCIENCES, 115478, MOSCOW, KASHIRSKOYE CHAUSSE 26, USSR

1. INTRODUCTION

Some years after vitamin C was isolated and characterized as L-ascorbic acid (AA), it was observed that the content of AA in cabbage increased when cabbage was boiled for a short time[1]. It was supposed that the increase is due to the liberation of "bound AA"[2]. Soon it was found that the "bound AA" can be extracted from cabbage with organic solvents[3]. Sen Gupta and Guha gave the name ascorbigen to the substance, although they did not isolate it[4].

Ascorbigen was isolated from fresh juice of Savoy cabbage in 1957[5]. It is formed in damaged or in stored aging plant tissues from the mustard oil glucoside, glucobrassicin - (indol-3-yl)acetothio-S-β-D-glucopyranosyl)hydroxyimyl-O-sulfate (1). It produces under the action of *mirosinase* (indol-3-yl)methylisothiocyanate (2), which is transformed into 3-hydroxymethylindole (3a), the latter reacts with AA to yield ascorbigen (4a)[6,7,8]. The other alkaloids of the glucobrassicin group, e.g. neoglucobrassicin (1-methoxyglucobrassicin)[9] or 4-methoxyglucobrassicin[10], which have been identified in several plant tissues, are sources of substituted ascorbigens in plant extracts.

The ascorbigen content of plant extract depends on species, age, parts of plants, conditions of growth etc.

The concentration is especially high in different species of cabbage, e.g. growing kohlrabi tips contain 6.66 g/l of **4a**[11,12]. It was identified in many other vegetables and fruits, including potato tubers[13]. Additionally to **4a**, products of its transformation were found in plant extracts; these are ingredients of vegetable or fruit diet. The structure of compound $C(C_1)$, which accompanies **4a** in plant extracts, was not elucidated, but both **4a** and $C(C_1)$ were found in blood and urine of humans, who had obtained ascorbigen orally with food[14,15,16]. Some authors do not consider **4a** a natural compound but regard it as an artifact as it is produced during the isolation process[7,8]. This is a nomenclature question. At any rate **4a** was isolated from natural products, and as humans and animals get noticeable amounts of ascorbigen with food, it is reasonable to assume that these compounds participate in biological effects of vegetable or fruit diet at least as depot-forms of **AA**.

2. SYNTHESIS OF ASCORBIGENS AND OF THEIR ACETYL AND METHYL DERIVATIVES

Investigation of 2-C-benzylation of **AA** with benzyl chloride has helped to elucidate the **4a** structure[17]. In 1966 G. Kiss and H. Neukom demonstrated, that by

the interaction of **3a** and **AA** ascorbigen A, which is 2-C-
[(1-H-indol-3-yl)methyl]-β-L-threo-L-glycero-3-hexulo-
furanosonic acid γ-lactone, identical to natural ascorbigen, is formed. Additionally to **4a** small amounts of
ascorbigen B were isolated, which was supposed to be the
2-epimer of ascorbigen A, i.e. 2-[(1-H-indol-3-yl)
methyl]-β-L-threo-D-glycero-3-hexulofuranosonic acid
γ-lactone (**5a**)[18].

(5a)

3-Hydroxymethylindole **3a** and its analogues are
key-compounds in the synthesis of **4a** and its
congeners[19]. There are two routes to these compounds:
the direct hydroxymethylation of indoles (the A route)
and the reduction of available substituted 3-formyl-
indoles with $NaBH_4$ (the B route).

Hydroxymethylation of indole (**6a**) with paraform-
aldehyde in methanol in the presence of NaOMe gives **3a**
in quantitative yield, 5-bromoindole (**6i**) or 2-
methylindole (**6j**) give **3i** or **3j** respectively in lower
yields[20]. 1-Methylindole or 1-ethylindole do not
interact with formaldehyde under these conditions[20].

Compounds differently substituted in the indole
ring **3a-j,k** were obtained with good yields by the
reduction of the corresponding substituted 3-formyl-
indoles **7a-i,k** with $NaBH_4$[21,22]. A series of substituted
ascorbigens **4a-i,k** were prepared by the interaction of
3a-i,k with **AA** in ~50 % yields[19-22]. Di- or poly(indol-
3-yl)methanes are formed as by-products in these
reactions as well as minor ascorbigen-like compounds,
which at first were supposed to be ascorbigen B (**5a**)
analogues. Formation of di(indol-3-yl)methane from two
molecules of 3-hydroxymethylindole with the release of

a. R' = R'' = X = H   (A,B)
b. R' = CH$_3$; R'' = X = H   (B)
c. R' = C$_2$H$_5$; R'' = X = H   (B)
d. R' = n-C$_3$H$_7$; R'' = X = H   (B)
e. R' = n-C$_4$H$_9$; R'' = X = H   (B)
f. R' = CH$_2$CH=CH$_2$; R'' = X = H   (B)
g. R' = CH$_2$C$_6$H$_5$; R'' = X = H   (B)
h. R' = R'' = CH$_3$; X = OCH$_3$ (B)
i. R' = R'' = H; X = Br   (A,B)
j. R' = X = H (A); R'' = CH$_3$
k. R' = CH$_2$COOH; R'' = X = H   (B)

(For compounds 3 - 19))

formaldehyde[23], and its further 2-(indol-3-yl)methylation to produce polymeric compounds[24] were described previously.

N-Substituted ascorbigens **4b** and **4c** were obtained with ~ 10 % yields by the interaction of **6b** or **6c** with CH$_2$O and AA, the latter was used simultaneously as a catalyst of hydroxymethylation and as a trap for the intermediate **3b** or **3c**. For the first time this "one pot reaction" was used for the preparation of **4a**[25]; **4i** and **4j** were also prepared by this method with low yields[20].

In the ^{1}H-n.m.r.spectra of **4a-k** the value (J_{45}~0) of spin coupling constant demonstrates that the predominant conformer has the dihedral angle H-C_4-C_5-H close to 90°[21].

By the action of methanolic HCl ascorbigens **4a,b,i** produce the corresponding 3-O-methyl glycosides **8a,b,i**; acetylation of them gave 5-O-acetyl derivatives **9a,b,i**[21]. More bulky alcohols do not produce 3-O-alkylascorbigens. By the acetylation of **4a,b,i** with Ac_2O in pyridine corresponding 2,5-di-O-acetylascorbigens (**10a,b**); 2,3,5-tri-O-acetylascorbigens (**11a,b,i**) and 2,5-di-O-acetyl-3,4-anhydroascorbigens (**12a,b**) were obtained[21]. The ^{1}H-n.m.r. and ^{13}C-n.m.r. data are in accordance with the structures **10**, **11** and **12**. In the case of 5'-bromoascorbigen (**4i**) the indole NH group is also acetylated to give **11l** and **12l**[21].

(**8a,b,i**.R = H)
(**9a,b,i**.R = Ac)

(**10a,b**)

(**11a,b,i,l**)
l. R'= Ac; X = Br.

(**12a,b,l**)
l. R'= Ac; X = Br.

In i.r. spectra of **12a,b,l** the frequency of the OH group is absent, which excludes the alternative structure **13** for these compounds.

The CD spectra of **4, 8, 10** and **11** demonstrate a negative Cotton effect (CE) at 220 nm and positive CE's at 232-236 nm (strong) and at 270 nm (weak). The type of CD spectra for these compounds depends first of all on the chirality of the C-2 center, which is close to indole and lactone chromophores. Comparison of CD spectra of **9a, 10a** and **11a** demonstrates, that for **10a** and **11a** CE's are similar and two-fold lower then for **4a, 8a** and **9a**. This allowed to conclude that in di-O-acetylated compounds 2-OH and 5-OH groups are protected[21].

3. TRANSFORMATION OF ASCORBIGEN IN ACIDIC MEDIA; REVISION OF ASCORBIGEN B STRUCTURE

Ascorbigen can substitute vitamin C, but guinea-pigs are able to utilize only about 10-20 % of AA bound in ascorbigen[26,27]. At physiological pH and temperatures (pH 7.2, 37°C) **4a,b,c,g** form AA. After one hour the yield of AA from **4a** is about 24 %, **4b** and **4c** are even less stable in solution and release in these conditions about 40 % of AA. The rate of decomposition of **4** decreases with decreasing temperature[22]. I.p. administration of **4b** to mice produces a substantial increase of the AA level in mice serum[22].

The fate of the (indol-3-yl)methyl moiety during this transformation was not understood for a long period. However, we have found that the transformation of ascorbigen by heating at pH< 7 in water proceeds by an intermolecular mechanism involving release of AA and addition of (indol-3-yl)methyl cation to another molecule of **4a** to yield N'- [(indol''-3-yl)methyl]-ascorbigen (**15a**), 2'-[(indol-3''-yl)methyl]ascorbigen (**14a**) and then 2'-{2''-[(indol-3'''-yl)methyl]indol-3''-yl}methylascorbigen (**16a**), which were separated by t.l.c[28]. For example, after 3 h heating of aqueous solution of **4a** at 37°C and pH 5 **15a**, **14a** and **16a** were obtained in yields of 2 %, 15 % and 10.5 %, respectively.

These substances were characterized by polarimetric, ^{1}H-n.m.r., Fab-Ms, i.r. and u.v. data.

CD spectra of **4a**, **14a**, **15a**, and **16a** show positive CE at 235 nm thus verifying the epimeric (S) configuration at 2-C atom. We have not found any 2-C epimeric compounds among the products.

T.l.c. and HPLC show that **14a** is formed from **4a** at 37° - 60°C and pH 1 - 6, yields of **16a** being decreased with increasing pH ; the yields of **15a** in these conditions are very low (Figure 1). T.l.c.and HPLC also demonstrate, that at pH 7.2 and 37°C natural ascorbigen **4a** releases AA and produces poly(indolyl)methanes rather than **14a**, **15a** and **16a**. For addition 1-deoxy-1-(indol-3-yl)-α-L-sorbopyranose (see below) is formed from **4a** via alkaline decarboxylation and isomerization.

T.l.c. and HPLC methods show that ascorbigen B[18,29] is the mixture of **15a**, **14a** and **16a**, with **14a** prevailing.

Similar transformations were performed for 5'-bromoascorbigen (**4i**) to give **14i** and **16i**. 1'-Methylascorbigen (**4b**) and 1'-butylascorbigen (**4e**) gave at 37°C and pH 5 **14b** or **14e**, respectively. No analogues of **16a** could be isolated.

HPLC demonstrates that compounds **4a**, **14a**, **15a** and **16a** are components of the ethyl acetate extract from fresh cabbage juice and from sauerkraut juice (Figure 2).

These results suggests that vegetables and fruits contain not only ascorbigen, **4a**, but also a family of related compounds and that under biological conditions the (indol-3-yl)methyl moiety from ascorbigen can be transferred to nucleophilic centers of other

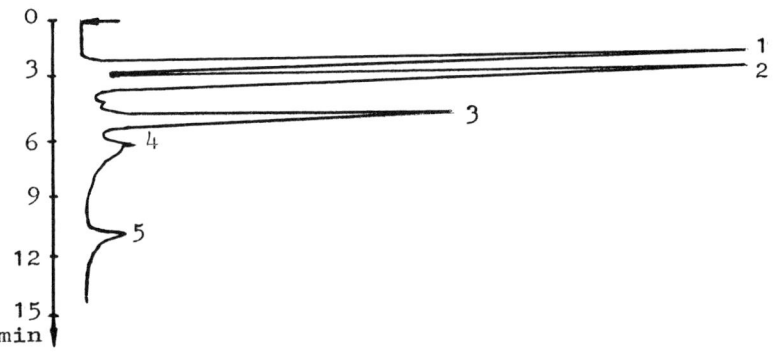

Figure 1. Chromatogram of 4a in solution after 2 h incubation at pH 5.0 and 60° C (Column 250 x 4.6 mm ID, Zorbax ODS, mobile phase 0.1 M ammonium acetate - acetonitrile pH 5.7 (60:40), flow rate: 1 ml/min. Peaks: 1 - AA, 2- 4a, 3 - 14a, 4 - 15a, 5 - 16a.

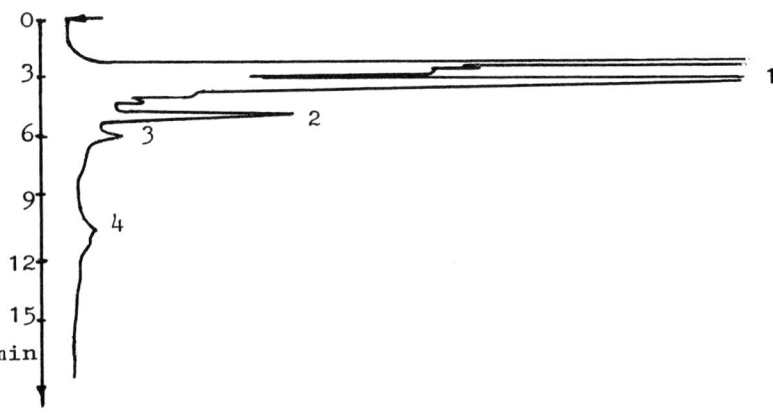

Figure 2. Chromatogram of ethyl acetate extract from sauerkraut juice. For HPLC conditions see Fig.1. Peaks: 1 - 4a, 2 - 14a, 3 - 15a, 4 - 16a.

biologically important molecules.

4. TRANSFORMATIONS OF ASCORBIGENS IN ALKALINE MEDIA

We have found a new transformation of ascorbigens, which is not connected with the release of AA[30]. In alkaline solutions at pH 9 **4a** opens the lactone ring, decarboxylates and rearranges to produce 1-deoxy-1-(indol-3-yl)-α-L-sorbopyranose (**18a**) with the yield of 30 %. Similar transformations were performed for **4b**, **4d**, **4e** and **4f**. The ion **17** is presumably an intermediate in this reaction. An analogous transformation was previously described for 2-benzyl-3-hexulofuranosonic acid γ-lactone, which has been found to produce 1-C-phenyl-2- hexulose, but the stereochemistry of this compound has not been elucidated[17]. Compounds **18** are not stable and decompose after 1-2 days of storage in refrigerator. Formation of (indol-3-yl)acetic acid and (indol-3-yl)carboxylic acid from ascorbigen under the prolonged action of alkali has been described [15].

^{1}H-n.m.r. spectra of **18** were compared with that of α-L-sorbopyranose (**19**). The signals from the carbohydrate moiety of **18** agreed nicely with those of **19**. The values of $J_{3,4}$, $J_{4,5}$ and $J_{5,6ax}$, which comply with trans-diaxial disposition of corresponding protons, demonstrate that compounds **18** and **19** have $^2C_5$ chair conformation. Upfield shift of 3-H and 5-H in comparison with **19** (Δδ ~0.2 ppm) and insignificant change in positions of 4-H and 6-H_{ax} signals suggests that in all cases the 2-OH group is axial, and the compounds **18a,b,d,e** and **f** have α-configuration. We succeeded in isolation of the individual 3,4,5-tri-O-acetyl-1-deoxy-1-(1-butylindol-3-yl)-α-L-sorbopyranose (**20**), when (**18e**) was acetylated in pyridine at -40°C with acetic anhydride.

All ascorbigens **4** in solution at pH>7 produce slowly the corresponding 1-deoxy-1-(indol-3-yl)-

α-l-sorbopyranoses, **18**. The possibility of formation of **18a** from **4a** in natural products or in animal or human bodies cannot be excluded.

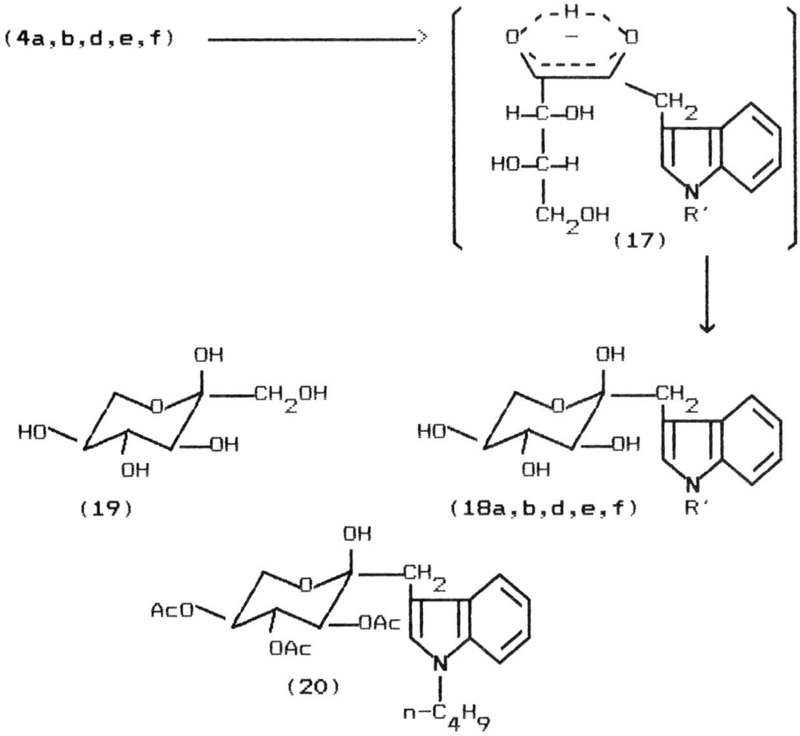

(4a,b,d,e,f) ⟶ [structure **(17)**]

↓

(19)   (18a,b,d,e,f)   R'

(20)   n-C$_4$H$_9$

5. INTERACTION OF ASCORBIGENS WITH AMINES

Interaction of **4a** or **4i** with ammonia in methanol at room temperature produces 2-C-[(indol-3-yl)methyl]-β-L-threo-L-glycero-3-hexulofuranosonic acid amide (**21m**) or its 5'-bromoderivative **21v**[18,21]. The formation of amide is not accompanied by the opening of the furanose ring, but the conformation of the furanose cycle in comparison with that of the corresponding ascorbigen is changed[21]. Similarly, ascorbigens **4a,b,d,e** and **f** interact with primary alkylamines or arylalkylamines to give amides **23n-u**. 5-Methoxytryptamine, isopropylamine, benzylamine and 2-hydroxyethylamine were used in these transformations. Amides **21** are not stable at room temperature.

In ^{1}H-n.m.r. spectra of compounds **21** signals of 6-Ha and 6-Hb are doublets of doublets, as for compounds **4**, thus providing an argument for cyclic 5-membered structures. The proton 4-H is shifted downfields (in comparison with the corresponding ascorbigens) due to the unsubstituted 4-OH group (for **21k** $\Delta\delta$ ~0.5 ppm). Acetylation of amides **21m** or **21u** yielded the acyclic 2,4,5- tri-O-acetylderivatives **22m** or **22u**, respectively[21,31]. In ^{1}H-n.m.r.spectra of these compounds there are three acetoxy groups; as the signals of the protons 4-H and 5-H are shifted downfields ($\Delta\delta$ 1.5 - 2.5 ppm), but the positions of 6-Ha and 6-Hb signals are not changed, the acetoxy groups are in positions 2, 4 and 5. The opening of the furanose ring occurs after acetylation, because the primary hydroxy group is not substituted. The acyclic structure of acetylated amide **22m** was confirmed by ^{13}C-n.m.r. data[21].

Arylamines, for example aniline or p-toluidine, do not react with ascorbigens in methanol at room temperature even during several days, which can be explained by their lower basicity. The possibility to interact with amines is remarkably dependent on sterical availability of amino groups. T.l.c. demonstrates, that the interaction of **4b** with 1-(indol-3-yl)-2-amino-propane at room temperature proceeds much more slowly than the interaction with 5-methoxytryptamine or isopropylamine and additionally to formation of **21s**, which was isolated by t.l.c. method with the yield of 10 %, concurrent decarboxylation and formation of sorbose derivative **18b** took place. At room temperature even after one month not all starting material (**4b**) was consumed. We have not succeeded in obtaining products of interaction of **4b** with 6-aminopenicillanic acid (unblocked or per-trimethylsilylated form); with the base of antibiotic daunorubicin, interaction of **4b** with kanamycin base produced sorbose derivative **18b**.

(**4a,b,d,f,i**) —RNH$_2$→ (21m-v)

(22 m,u)

m. R' = R = X = H
n. R' = X = H; R = CH$_2$CH$_2$Ind'
o. R' = CH$_3$; X = H; R = CH$_2$CH$_2$Ind'
p. R' = CH$_3$; X = H; R = CH(CH$_3$)$_2$
q. R' = CH$_3$; X = H; R = CH$_2$C$_6$H$_5$
r. R' = CH$_3$; X = H; R = CH$_2$CH$_2$OH
s. R' = CH$_3$; R = CH(CH$_3$)CH$_2$Ind
t. R' = n-C$_3$H$_7$; X = H; R = CH(CH$_3$)CH$_2$Ind
u. R' = CH$_2$CH=CH$_2$; X = H; R = CH$_2$CH$_2$Ind'

Ind= (indole)
Ind'= (methoxyindole, -OCH$_3$)

Secondary amines, for example morpholine (pK 8.7) as well as phenylhydrazine do not interact with ascorbigens at room temperature, while piperidine (pK$_a$ 11.2) produces the sorbose derivative **18b**.

6. DERIVATIVES OF ASCORBIGENS AS SHOT-GUN DRUGS

As ascorbigens in mild conditions nonenzymatically disintegrate to **AA** and various compounds containing (indol-3-yl)methyl residue and also products of decarboxylation and ascorbigen amides, they can be used in the design of compounds, which we propose to call "shot-gun drugs". If a biologically active compound linked with the ascorbigen molecule through the indole ring is given to an animal, not only this ascorbigen derivative will be present in the body, but also a series of products of its transformation plus AA, interacting with wider variety of molecular targets, than the parent compound does. It allows to consider drugs linked with ascorbigens, as "shot-gun drugs". These compounds may differ from parent compounds in distribution, transport to targets, pharmacokinetic and other parameters, thus leading to multiple mechanisms of action modulated by high level of AA.

The first representatives of shot-gun drugs were synthesized from 1'-carboxymethylascorbigen (**4k**), which was prepared from 1-carboxymethyl-3-hydroxymethyl-indole (**3k**) and AA. N-Hydroxysuccinimide ester **23** was obtained from **4k**, whereupon it was conjugated with antitumour antibiotics daunorubicin, carminomycin or doxorubicin to give **24a,b** or **c** respectively. The antitumor properties of these compounds are now under investigation. Studies of shot-gun drugs with different spacers between ascorbigens and biologically active compounds is now in progress.

7. BIOLOGICAL PROPERTIES OF ASCORBIGENS

Cytostatic and antitumour activity

Ascorbigens do not demonstrate cytostatic, antibacterial or antiviral activity *in vitro*, but some of them inhibit growth of murine experimental tumours *in vivo*. 1'-Methylascorbigen (**4b**) (50 - 75 mg/kg during 5 days i.p.) has the most pronounced antitumour effect on murine adenocarcinoma Ca 755 (T/C 79-84 %), Lewis lung tumor (74 - 76 %), tumour of large intestine AKATOL (50 - 60 %), sarcoma 37 (52 - 63 %) and leukemia P 388 (ILS 37 %). In control experiments **AA** demonstrated marginal activity against tumour Ca 755 (~ 50 %), AKATOL (~50 %) at the doses 25 - 100 mg/kg during 5 days^{21}. Ascorbigens **4a**, **4c**, and the 3-hydroxymethylindoles **3a** and **3b** had no antitumour activity. It is suggested that the antitumour activity of **4b** is related to its immunomodulatory properties^{21}.

Immunomodulatory properties of ascorbigens

In vitro ascorbigens **4a-h** do not demonstrate pronounced immunomodulatory activities. They decrease NK cells cytotoxicity, stimulate at low concentrations and inhibit at high concentrations T lymphocyte blastogenesis induced by con A, they do not significantly influence the formation of cytotoxic T lymphocytes in allogenic mixed lymphocyte culture, but they augment cytostatic activity of macrophages against leukemia P 388 cells.

Administration of some of the ascorbigens to animals leads to pronounced biological effects; the most potent immunomodulator among all ascorbigens tested is 1-methylascorbigen (**4b**). In mice after i.p. injection it enhances the cytostatic activity of macrophages, the activity of NK cells, the generation of cytotoxic T lymphocytes in allogenic mixed lymphocyte culture and slightly decreases the number of antibody producing cells (in Jerne hemolytic plaque assay)^{32,33}. **4b** significantly affects the metabolism of arachidonic acid in mice splenocytes with resulting increases of prostaglandin E_2 (up to 550 %) and leukotriene B_4 (up to 450 %) production during the first hours after administration.

7. CONCLUSION

The transformation of **AA** into ascorbigen during cell damage represents a process of **AA** conservation when it is protected from air oxygen action. Natural ascorbigen **4a** is biologically less active than its analogue **4b**, but the biological role of **4a**, which is eaten by humans and animals remains to be discovered.

If biological effects of ascorbigens, including **4b** are dependent only on their ability to release **AA**, the differences in their biological properties *in vivo* can be related to the differences in the concentrations of these compounds in the specific organs and tissues, which can be dependent on their lipophilicity.

We cannot also exclude the possibility of **4b** bioactivation either.

The release of **AA** from ascorbigens is accompanied by production of (substituted) indolylmethyl cations, which can attack nucleophilic centers of various important molecules in cells. These alkylation reactions may contribute to the biological effects of ascorbigens.

It will be interesting to evaluate the biological properties of ascorbigens **14**, **15** and **16**, as well as di- and poly(indolyl)methanes and other products of ascorbigen transformations including sorbose derivatives 18 and amides 21.

At the same time there is a possibility, that the biological properties of ascorbigens are connected primarily to intrinsic properties of the ascorbigen structure and only after this with the release of AA or formation of indole-containing products from ascorbigen transformation. Here we have to mention the recent work of G. Fodor and collaborators[34-38]. These authors synthesized 2-C-alkylated AA compounds, which have structural elements similar to ascorbigen but without the indole moiety (25a,b,c). Some of them are immunomodulators as they enhance mitogen induced T lymphocytes blastogenesis and the production of inter- leukin 2 and interferon. No data about the ability of compounds 25a,b,c to release AA have been published.

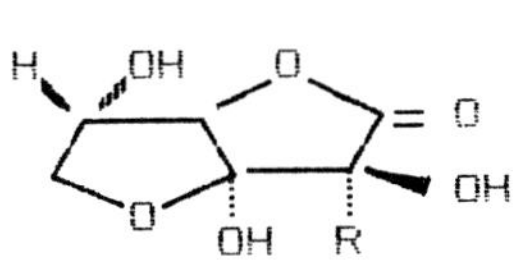

(25a,b,c)

a.

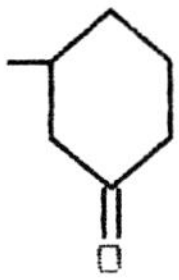

b. $-CH_2CH_2COCH_3$

c.

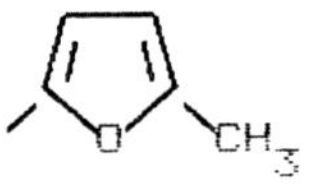

Further studies will elucidate the biological role of natural ascorbigen and the molecular basis of the biological activity of 1'-methylascorbigen and other AA 2-C-alkylderivatives.

REFERENCES

1. B. Ahmad, Biochem. J., 1935, 29, 275.
2. E.W. McHenry and M.L. Graham, Nature, 1937, 135, 871.
3. B.C. Guha and J.C. Pal, Nature, 1936, 137, 946; 1937, 139, 844.
4. P.N.Sen Gupta and B.C. Guha, Nature, 1938, 141, 974
5. Z. Prochazka, V. Sanda, F. Sorm, Coll. Czech. Chem. Commun., 1957, 22, 333, 654.
6. R. Gmelin and A.I. Virtanen, Suomen Kemistilehti, 1961, 34B, 15.
7. R. Gmelin and A.I.Virtanen, Ann. Acad. Sci. Fenniciae. Ser. A. II. Chemica. 1961, # 107, 1.
8. M. Kutacek, Z. Prochazka, and K. Veres, Nature, 1969, 194, 393.
9. R. Gmelin and A.I.Virtanen, Acta Chem. Scand., 1962, 16, 1378.
10. J.W. Truscot, I.R. Minchinton, D.G. Burke, J.P. Sang, Biochem. and Biophys. Res. Commun., 1982, 107, 1368.
11. M. Kutacek, M. Valenta, and F. Icha, Experientia, 1957, 13, 284.
12. C. Kumar and N.J. Chandrasekhara, Food Sci.Technol., 1973, 10, 42.
13. W. Feldheim, Ind. Obst. Gemuesenverwert., 1970, 55, 27.
14. W. Feldheim., Z. Prochazka., Intern. Z. Vitaminforsch., 1962, 32, 251.
15. V. Sanda, N. Constantzas, and Z. Prochazka, Coll. Chem. Commun., 1962, 27, 554.
16. H. Schmandke, Lebensmit. Ind., 1967, 14, 25.
17. K.G.A.Jackson and J.K.N. Jones, Can. J. Chem., 1965, 43, 450.
18. G. Kiss and H. Neukom, Helv. Chim. Acta, 1966, 49, 989.
19. A.I. Virtanen and E. Piironen, Suomen Kemistilehti, 1962, 35B, 104.
20. I.L. Plikhtyak, I.V. Yartseva, N.Y. Podkhaliusina, and M.N. Preobrazhenskaya, Khim. Geterocykl. Soedin. in press.
21. V.I. Mukhanov, I.L. Yartseva, B.S. Kykot, Y.Y. Volodin, I.L. Kustova, N.A. Lesnaya, Z.P. Sof'ina, N.P. Ermakova and M.N. Preobrazhenskaya, Bioorg. Khim. 1984, 10, 544.
22. Y.V. Bukin, I.L. Plikhtyak, V.A. Draudin-Krylenko,I.V. Yartseva, L.M. Orlova, and M.N. Preobrazhenskaya, Bioorg. Khim., 1987, 13, 539.
23. R. Leete, J. Am. Chem. Soc., 1959, 81, 6023.
24. F. Amat-Guerri, R. Martinez-Utrilla and C. Pascual, J. Chem. Res.(S), 1984, 160.
25. A.I. Virtanen and M. Kiesvaara, Acta Chem. Scand, 1963, 17, 848.
26. M. Kiesvaara and A. Virtanen, Acta Chem. Scand., 1963, 17, 849.
27. K. Matano and N. Kato, Acta Chem. Scand., 1967, 21, 2886.

28. M.N. Preobrazhenskaya, A.M. Korolev, E.I. Lazhko, L.G. Aleksandrova, I.L. Plikhtyak, I.V. Yartseva, J. Chem. Soc. Chem. Commun., in press.
29. The Merck Index. 10th Edition. Merck & Co., Inc. Rahway, N.Y. USA. 1983, # 847, p. 120.
30. I.L. Plikhtyak, I.V. Yartseva, N.A. Kluev, and M.N. Preobrazhenskaya, Khim. Geterocykl. Soedin., 1984, 607.
31. I.L. Plikhtyak, I.V. Yartseva, Khan Zu En, and M.N. Preobrazhenskaya, Bioorg. Khim., 1988, 14, 1437
32. S.A. Efimov. Antibiotiki i Khimioterapia, 1989, 125.
33. S.A. Efimov, V.E. Shevchenko, M.M. Viadro, I.A. Shchukin and M.N.Preobrazhenskaya. To be published.
34. R. Arnold, G. Fodor, C. George, I. Karle., Can. J. Chem., 1987, 65, 131.
35. K. Sussangkarn, G. Fodor, I. Karle, and C. George, Tetrahedron, 1988, 44, 7047.
36. G. Fodor, R.W. Veltri, PCT Int. Appl. WO 8706, 585 (C1C07D493/04) 05 Nov1987. US appl. 857,291 Apr. 1986. C.A. 1989, 109, 129601v.
37. R.W. Veltry, M.W. Baseler, G.B. Fodor, K. Sussangkarn, and P.E.Maxim. Fed. Proc. 1987, 46, 454.
38. G. Fodor, K. Sussangkarn, R. Arnold, H. Mathelier, T. Mohasci, R. Mujumdar, J. Butterick, and R.W.Veltri Acta Biochim. Biophys. Hung., 1987, 22, 165.

Some New Data on Structure, Properties, and Mechanism of Action of 1,2-Dihydro-3H-1,4-Benzodiazepin-2-ones

S. A. Andronati and T. A. Voronina

A. V. BOGATSKY PHYSICO-CHEMICAL INSTITUTE, ACADEMY OF SCIENCES OF THE UKRAINIAN SSR, ODESSA, 270080

Drugs used for the treatment of anxiety, aggressiveness and fear include benzodiazepines and related fused systems. In addition to the anxiolytic effect these drugs also exhibit hypnosedative and anticonvulsant properties.

Correlation of various properties of the drugs with their structure and geometry were made. This work lead to the preparation of new potentially active compounds. The character of the substituents and groups as well as the nature and site of junction of the annelated rings on the activity was qualitatively determined. The relationship between the geometric parameters of 1,2-dihydro-3H-1,4-benzodiazepin-2-ones obtained by means of X-ray analysis and the activity were also considered.

Attempts of quantitative estimation of the relationship between the activity, physicochemical characteristics (lipophilicity, pK, polarographic characteristics, dipole moments, atomic charges) and the structure of benzodiazepines have been generalized in a number of papers [1-3].

Application of the method of Hansch for the quantitative study of relationship between physicochemical and pharmacological properties of benzodiazepines gave satisfactory results only for rather narrow series of benzodiazepines. Apparently, for the design of novel anxiolytic, anticonvulsant and hypnosedative agents the results of the investigation of the mechanism of their action are of great importance, as well as the results of structure-activity relationship studies for the agonists and antagonists of the appropriate receptors.

The problem of the mechanism of action of 1,4-benzodiazepines and related systems has attracted the interest of researchers for the last 20 years. Since the end of the sixties the effect of benzodiazepines on the neuromediator systems has been thoroughly studied. It was shown that they alter the content and rate of metabolism of γ-aminobutyric acid (GABA), glycine, dopamine, 5-hydroxytryptamine, norepinephrine and acetylcholine in the nerve tissue. An important result of these consisted in the establishment that benzodiazepines potentiate GABAergic processes and that the mechanism of their action is connected with the facilitation of postsynaptic GABAergic regulation.

The most significant step in the cognition of physicochemical foundations of the central nervous system (CNS) activities was the discovery of the opiate receptors by Pert and Snider in 1973 followed by the discoveries of a series of psychoactive peptides. During that period the modern efficient methods were worked out for the investigation of processes in CNS on the molecular level which promoted further investigations of the nervous system receptors.

The striking discovery was made in 1977, when the highly specific sites of binding synthetic psychotropic agents benzodiazepines were found on synaptosomal membranes of experimental animals and human brain. This discovery is as important for the physicochemical biology of CNS as the discovery of the opiate receptors

In spite of the vast investigations concerning benzodiazepine receptors (BDR) [4] a number of questions remain.

It is known that BDR allosterically interact with GABA receptors (GABAR), with the binding sites of barbiturates and picrotoxin, and the chloride channel. In some publications it is noted that protein molecules containing GABA and benzodiazepine (BD) receptors are not identical. The different molecular weights of the corresponding proteins also substantiate this suggestion. However, at the purification of BDR during attempts to separate the binding sites of GABA and BD were not successful.

Another controversial problem is connected with the BDR heterogeneity. The existence of two BDR populations, central and peripheric has been proven. The question is how many central benzodiazepine receptors (CBDR) really exist, which they are and how they function

Peripheric benzodiazepine receptors (PBDR) of man are naturally more accessible objects for investigation than CBDR. For instance, the former are present in blood cells thrombocytes, lymphocytes and monocytes It was shown that the change of CBDR state under the effect of various factors (substances, psychoemotional stress) correlates with the change of the PBDR state. If this is so, the possibility exists for a control of the CBDR state in man according to the state of PBDR. This is important both for studying novel psychotropic agents and for the clinical practice.

Further investigations of PBDR will hopefully establish the role of BDs in the treatment of disorders which are not directly connected with CNS (for example, gastric ulcer). Besides, the anxiolytic and other types of BD effects will be perhaps characterized more completely on the level of the organism. In particular, it can be suggested that the inhibition of the increase of corticosteroid levels in blood is due to their direct effect on adrenal glands.

The suggested responsibility of various BDR subtypes for the pharmacological effects of BD is interesting since the possibility appears to prepare selective psychotropic agents. However, as noted previously, there is no understanding of CBDR heterogeneity and the suitability of this approach for the synthesis of selective preparations has not yet been proven.

It should be noted that depending on the experimental

conditions of CBDR isolation different (sometimes even contradictory) results can be obtained. Thus, one should critically analyze the data obtained by different authors.

Binding of benzodiazepines with BDR is characterized by reversibility, fast kinetics specificity (including stereospecificity), limited binding sites; GABA potentiation benzodiazepine binding of BDR.

In the literature the high specificity of BDR has been noted. More than 200 compounds belonging to various pharmacological classes did not show marked affinity for BDR. However, compounds of non-benzodiazepine structures possessing both BD agonist and antagonist properties were found.

The results of CBDR studies may be used for search of novel psychotropic agents in three directions.

1. Synthesis of novel BDR agonists and antagonists on the base of BDR ligand differentiation;
2. Synthesis of analogs of hypothetic endogenous BDR ligands;
3. Synthesis of novel compounds on the basis of the quantitative study of "structure-affinity for BDR-activity" relationship.

Let us now consider a few examples of these approaches.

BDR Agonists

BDR agonists are compounds competitively binding to BDR and possessing benzodiazepine-like properties. They include [1,4]benzodiazepines, I, related heterocycles, II-III, triazolobenzodiazepines, IV, triazolopyridazine, V, nicotinamide, VI, inosine, VII, hypoxanthine, VIII, zopiclon, IX, suriclon, X, and the substituted β-carboline-3-carboxylic acid ester, XI.

BDR Antagonists

BDR antagonists (imidazobenzodiazepines XII-XIV, pyrazoloquinoline XV, and the preparation ZK 93426 XVI), are compounds binding to BDR, not possessing any benzodiazepine-like properties; they block the pharmacological and biochemical effects of BDs and β-carbolines.

Inverse Agonists

They bind to the BDR and show inverse effects as compared with BD. This group includes the esters of β-carboline-3-carboxylic acid XVII and XVIII.

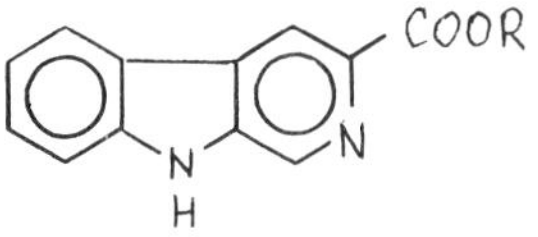

XVII

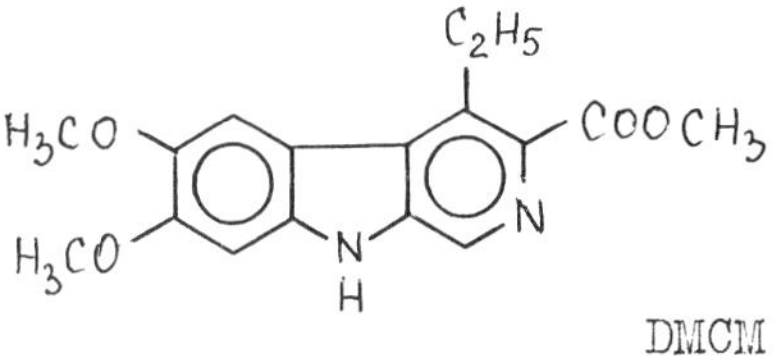

XVIII DMCM ZK 90798

Partial Antagonists

Partial antagonists have affinity to the BDR and show impaired few BD properties. They include β-carboline and pyrazoloquinoline derivatives - ZK 91926 (XIX) and CGS 9896 (XX) possessing anticonvulsant properties

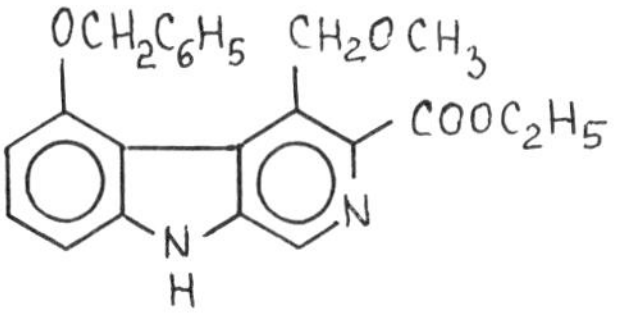

XIX ZK 91296

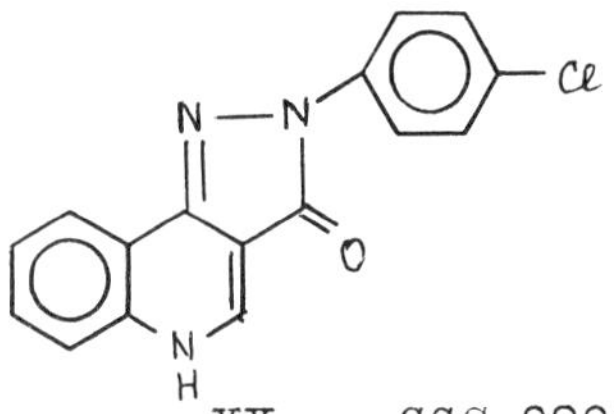

XX CGS 9896

Partial Inverse Agonists

They will bind to the BDR and possess anxiogenic or proconvulsant properties. β-Carboline-3-carboxylic acid methylamide XXI is an example of this group.

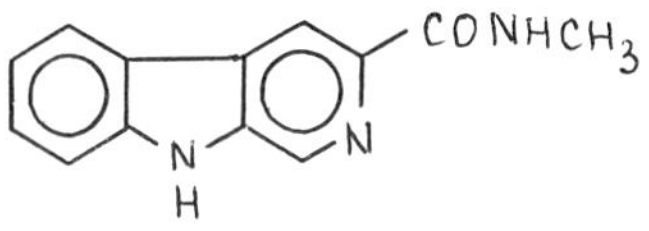

XXI FG 7142

Receptors are known to possess the corresponding endogenous ligands. However, the problem of the endogenous ligand for BDR has not yet been solved. The likely ligands are derivatives of β-carboline-3-carboxylic acid XV. The possibility of a peptide ligand has also been discussed, It is suggested that the polypeptide DBI (104 amino acid residues) extracted from rat and human brain can be its precursor.

Synthesis of analogs of hypothetic endogenous BDR ligands has resulted in a number of interesting psychopharmacologically active compounds. Thus among the novel harman derivatives promising anxiolytics, antidepressants and neuroleptics were found. 5-Methylbenzodiazepines, depending on the nature of the substituent R, these compounds display

$R = NO_2$, Br, H

XXII

different binding properties towards the receptor-ionophore ensemble. Thus, only the nitro derivative do possess some affinity for BDR. Pharmacologically it acts as a sedative and anticonvulsant agent. This compound does not influence the GABA-stimulation of diazepam binding to BDR.

Compounds XII (R=Br) and XII (R=H) suppress GABA-stimulation; they do not compete with diazepam for the binding with BDR. XII (R=H) behaves as a convulsant, its activity reaching that of picrotoxin. XII (R=Br) elicits aggressiveness of the experimental animals.

According to one of the versions about the physiological functionality of BDRs it is suggested that they are the "fear receptors", while BD act as the antagonists for these receptors.

The available data indicate that variations of the structure of 1,2-dihydro-3H-1,4-benzodiazepines may result in the formation of both BDR agonists and antagonists.

Studying the relationship between structure and affinity to BDR as well as the activity of dihydrobenzodiazepinenes and related heterosystems we have shown that 5-aryl-benzodiazepinones, 4-aryl-quinazolines, 6-aryl-1,5-benzodiazocinones and 7-aryl-1,6-benzodiazoninnones possess activity at BDR; a correlation was found between K_i values and activities of the different BD compounds (Fig. 1) [7].

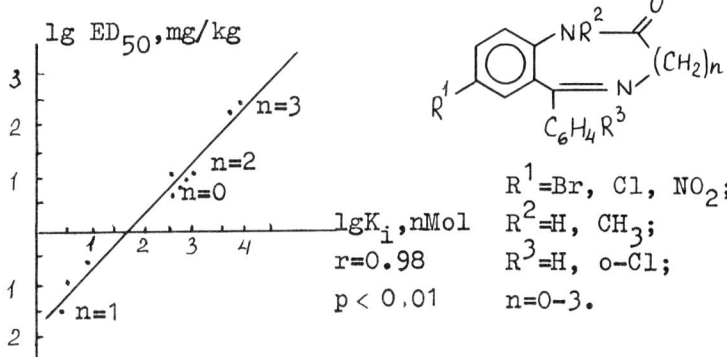

Fig. 1 Relationship between the affinity to BDR and the pharmacological activity *versus* pentylenetetrazol antagonism of the test compounds.

The data on binding of the different compounds to the receptors are especially important for screening and molecular design of novel drugs. In this case the influence of metabolism and pharmacokinetics on their final pharmacological characteristics is eliminated. However, in some cases there can be complications even with approach.

We have found [4], for example, that 7-difluoromethylthio-benzodiazepinone in experiments (in vitro) at the contact with the microsomal membranes of rat brain undergoes transformation to a 7-mercaptoderivative which can bind strongly to the protein. It seem likely that this transformation is responsible for the deviation seen in Fig. 2 [7].

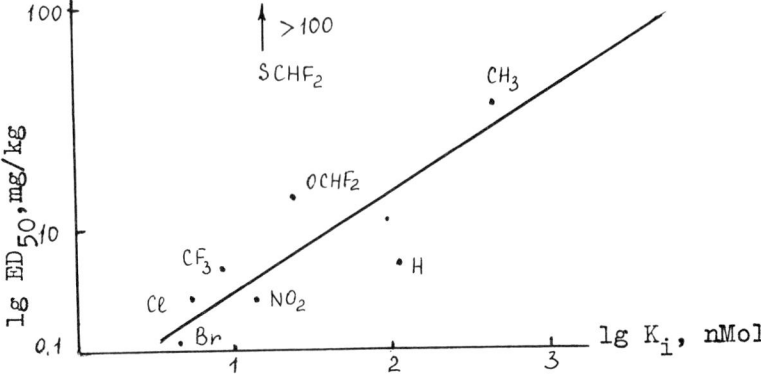

Fig. 2. Relationship between anticonvulsant activity of 7-substituted 5-phenyl-1,2-dihydro-3H-1,4-benzodiazepine-2-ones and their affinity to BDR.

Radioligand assays data and the results of a study of structure-activity relationship, as well as some literature data suggest that the benzophenonimine fragment is the basis of the BD pharmacophore and some BDR related ligands.

One of the most simple condensed heterocyclic systems containing such a fragment is 3-arylindazole, which indeed is able to compete for the BD binding sites.

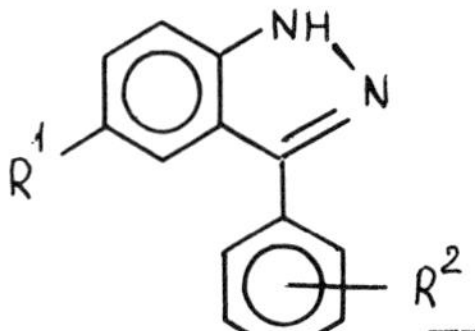

XXIII

$K_i = 33 \mu M$
$R_2^1 = Br$
$R^2 = H$

Affinity to BDR is exhibited by 1,2-dihydro-1,3-4-benzotriazepin-2-ones (XXIV), 1,2,3,4-tetrahydro-1,4,5-benzotriazocin-2-ones XXV, and acyclic synhydrazones of 2-syn-aminobenzophenones XXVI.

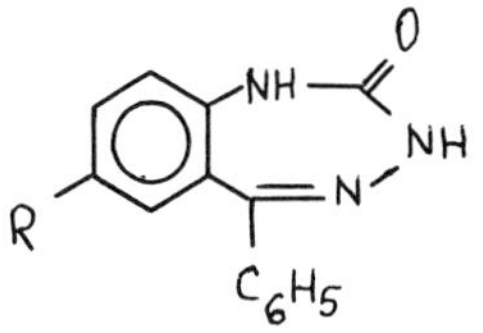

XXIV 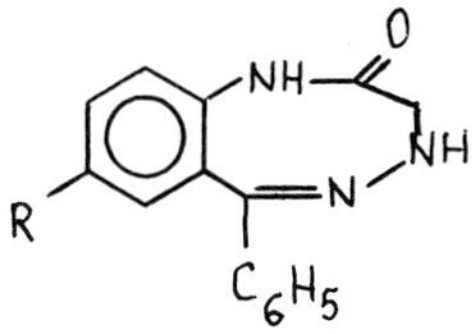   XXV 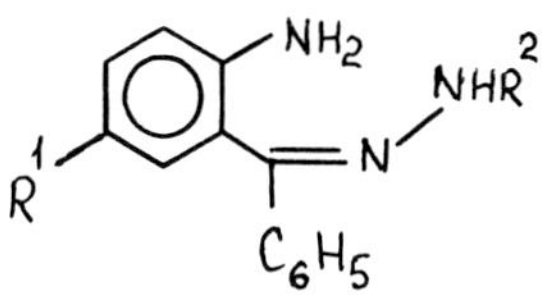   XXVI

As is well known, slight differences in steric structure are sometimes important for the "substrate-receptor" interaction. Thus, it is only natural that the questions of BD stereochemistry have attracted the interest of researchers for more than 20 years.

In the pioneering studies concerning the geometry and intramolecular mobility of benzodiazepine [8, 9] it was noted that 1,2-dihydro-3H-1,4-benzodiazepin-2-one molecules have the pseudo-boat conformation which can undergo inversion.

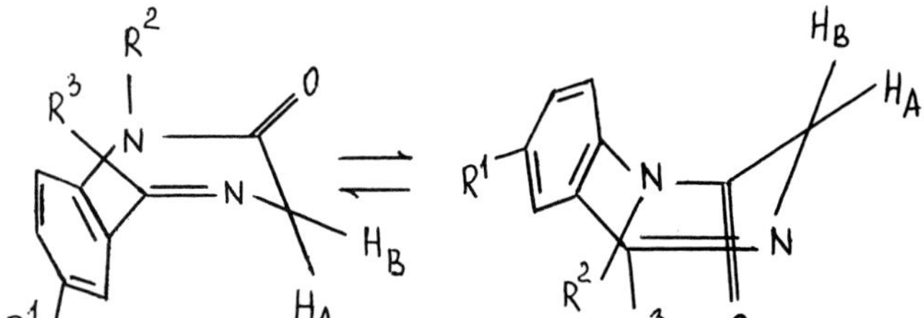

XXVII   XXVIII

By means of the dynamic NMR method the kinetics of the inversion of the 1,4-benzodiazepine derivatives was studied. The influence of their structure on the activation parameters for this process was considered [1].

The predominance of the pseudo-boat conformation of 1,2-dihydro-3H-1,43-benzodiazepin-2-ones in solution was confirmed by IR spectroscopy and dipole moments method.

In a series of investigations on the crystalline and molecular structure of 1,2-dihydro-3H-1,4-benzodiazepin-2-ones XXIX it was found that all the studied compounds in the solid phase possess the boat conformation. Comparison of the angles of the aryl substituent plane relative to the plane of the benzene ring with the pharmacological activity of 5-arylbenzodiazepinones did not reveal any "physiologically active conformations" of these compounds, that could be characterized by any optimal value of this angle [1].

XXIX

Thus, in the mentioned papers the pseudo-boat conformation for 1,2-dihydro-3H-1,4-benzodiazepin-2-ones has been verified both in the crystalline state and in solutions by X-ray as well as spectral studies. However, the following two points need further discussion.

$\underline{1}$. As a rule, practically no attention has been paid to the differences of the torsion angles characterizing the degree of the boat conformation distortion.

$\underline{2}$. The majority of the authors have dealt with the geometry and pharmacologically active compounds. However, we consider it more logical to compare compounds with similar structures and different properties.

On the base of these considerations and using various methods, we have studied the structure and properties of 1,2-dihydro-3H-1,4-benzodiazepin-2-ones of the formula XXX.

Some of the compounds of this series are highly active, while the activity of others is low.

IR, UV, photoelectron (PE), ^{1}H-NMR and mass spectra of electron resonance capture (ERC) were studied.

Analysis of mass spectra of ERC and PE spectra of the compounds at different temperatures proved that in gaseous phase their molecules exists in two unequivalent equilibrium forms C_1 and C_2 [10]. The curves of effective ion formation

(CEF) show splitted peaks. Owing to energy overlap, the CEF lines are poorly resolved (Fig. 3).

1. R^1=Br, R^2=o-Cl, R^3=R^4=H;
2. R^1=Cl, R^2=o-Cl, R^3=R^4=H;
3. R^1=Br, R^2=R^3=R^4=H;
4. R^1=OCHF$_2$, R^2=R^3=R^4=H;
5. R^1=Br, R^2=m-Cl, R^3=R^4=H;
6. R^1=Br, R^2=p-Cl, R^3=R^4=H;
7. R^1=CH$_3$, R^2=R^3=R^4=H;
8. R^1=Br, R^2=p-CH$_3$, R^3=R^4=H;
9. R^1=SCHF$_2$, R^2=R^3=R^4=H;
10. R^1=CF$_3$, R^2=R^3=R^4=H;
11. R^1=NO$_2$, R^2=R^4=H, R^3=CH$_3$;
12. R^1=NO$_2$, R^2=H, R^3=R^4=CH$_3$.

XXX

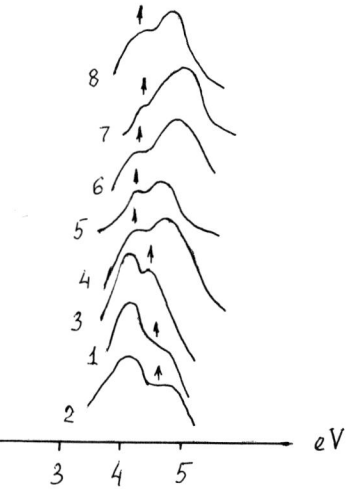

Fig. 3. CEF of M-H ions (compounds 1-8) at 180 °C. The arrows mark the increase of the relative intensities of the corresponding ions at this increased temperature.

At varying temperatures the position of the CEF peaks remain, while the intensities ratio are changing, so that the peak which is maximal at one temperature resolves poorly at another.

Extrapolation of the temperature dependence to low temperatures suggests that at room temperature the form C_1 is energetically more favorable for the compounds 1-3, while the form C_2 predominates for the compounds 6-8.

A similar relation is observed from the PE spectra of the compounds at different temperatures.

Thus, the data of ERC mass spectrometry and PE spectroscopy prove that in the gaseous phase 1,2-dihydro-3H-1,4-benzodiazepin-2-ones exist as two equilibrating forms (conformers) with unequivalent electron structures.

The conformers C_1 and C_2 may differ either in the diazepine ring geometry or in the value of the twisting angle of the aryl substituent at position 5 relative to the plane of the benzene ring.

In order to distinguish between the two conformers we have studied the temperature dependence of ERC mass spectra and PE spectra of the benzodiazepinones XXXI (R-H) and XXXI (R=CH_3) and the 4-phenylquinazolin-2-ones XXXII

XXXI
R=H, CH_3

XXXII
R=Br, $OCHF_2$, $SCHF_2$

The characteristics of the ERC mass spectra and PE spectra at different temperatures were observed in the spectra of the benzodiazepin-2-ones but were absent at the spectra of the quinazolines.

Since the conformational differences of the benzodiazepines XXXI can only be due to the difference in the diazepine ring geometries, and in the case of the quinazolines XXXII only rotamers are possible, we have concluded that 1,2-dihydro-3H-1,4-benzodiazepin-2-ones conformers differ by the geometry of the diazepine ring.

Analysis of benzodiazepine models suggests the possible existence of two conformers of the 7-membered ring: a pseudo-boat form with the cis-conformation of the amide group, and the flattened form with anticlinal conformation of the amide group.

The major difference between these forms lies in the dihedral angle φ between the planes containing the groups CNH and N-C=O.

XXXIII   XXXIV

The first conformation corresponds to the known pseudo-boat conformation of 1,4-benzodiazepines. The conclusion of the existence of the other conformer was suggested during the present investigation.

Quantum-chemical calculations of the conformers XXXIII and XXXIV by means of the MNDO method using the geometric parameters known from the X-ray structural analysis of 1,4-benzodiazepines proved the possibility of the existence of the second conformer - a flattened form with the anticlinal conformation of the amide group. The full energies of the first and second conformers differ by 0.5 eV.

Now the question arises if the different conformers of 1,2-dihydro-3H-1,4-benzodiazepines exist not only in the gaseous phase but also in solution and in crystalline state. The possibility of existence, resolution and identification of different 1,2-dihydro-3H-1,4-benzodiazepine conformers was demonstrated in 1983 [12]. The authors isolated two conformers of 2,2-dimethyl-5-phenyl-1,2-dihydro-3H-1,4-benzodiazepin-3-one in crystalline states. They had different dihedral angles in the diazepine ring. The angles of the phenyl ring rotation relative to the benzogroup were also different.

We suggest that we have obtained the indirect corroboration of the existence of stable 1,2-dihydro-3H-1,4-benzodiazepin-2-ones conformers in solution and solid phase.

Benzodiazepin-2-ones unsubstituted in position 1 are known to exist in solution as monomers, linear associates and cyclic dimers [1]. At the concentration below 2.5×10^{-3} mol/l there are practically no linear associates in CCl_4 solution. The IR spectra of solutions in CCl_4 (concentration 10^{-4} mol/l) of benzodiazepines unsubstituted in position 1 have the stretching vibrations of the carbonyl groups appearing as three partially overlapped absorption bands.

Thus, in the spectrum of compound 6 there is the absorption band of the carbonyl group of monomeric molecules at 1710 cm^{-1}, dimeric associates at 1690 cm^{-1}. At 1730 cm^{-1}, there is a third absorption band which is due to Fermi resonance. It remains for 1-substituted compounds at temperatures up to 70 °C. Thus, its appearance can not result from the formation of intermolecular associates. Rather it seems to be related to the absorption of the carbonyl group of the second conformer.

If it is supposed that unsubstituted benzodiazepinones are present in solutions as two conformers, and their ratio depends on temperature. Hence the ratio of the intensities of the C=O absorption bands of the two conformers and the cyclic dimers should also depend on temperature.

In fact, at different temperatures the redistribution of these band intensities was observed (Fig. 4).

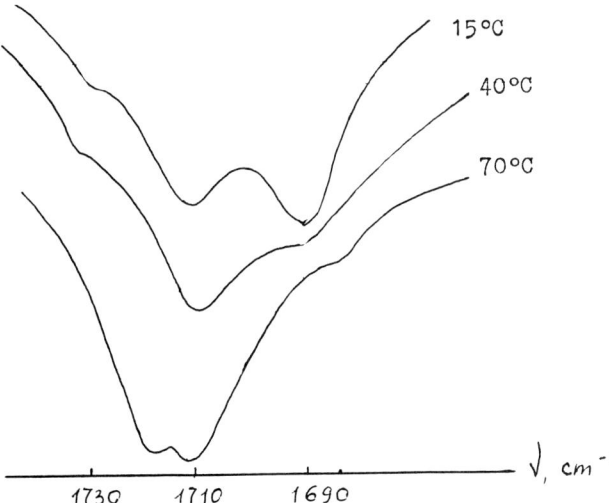

Fig. 4. Absorption band's of the stretching vibrations of the band C=O in CCl_4 (10^{-4} mol/l) at different temperatures.[4]

A similar situation is observed in the case of the absorption bands of NH-groups of monomeric and cyclically dimerized molecules of benzodiazepines unsubstituted in position 1.

The geometry of the 1,2-dihydro-3H-1,4-benzodiazepin-2-ones (XXXV) in crystalline state was studied using X-ray structural analysis [13].

$R=CH_3, C_2H_5, C_3H_7, i-C_3H_7, C_4H_9$

XXXV

All these compounds, like the 1,4-benzodiazepines described earlier, have the pseudo-boat conformation. However, the considerable differences in the boat distortion measured by the asymmetry index should be noted

$$G_s = \left[1/5(T_1^2 + T_4^2 + T_6^2 + (T_2+T_3)^2 + (T_5+T_7)^2 \right]^{1/2}$$

where T_1-T_7 denote torsion angles of the seven-membered ring, at N_1-C_2, C_2-C_3, C_3-N_4, N_4-C_5, C_5-C_{11}, $C_{11}-C_{10}$, $C_{10}-N_1$, respectively.

The value of G_s varies within the limits 1.3-9.1° for the series of compounds XXXV.

A comparison of the torsion angles of the amide group of various compounds XXXV and other described benzodiazepinones showed that the difference of the angle values is 23°.

It is interesting that Seiler and Zimmermann evaluating the twist-angles of 1,2-dihydro-3H-1,4-benzodiazepin-2-one conformations on the basis of UV spectral data obtained the values 27-51° [14].

The above information proved that both in solution and in solid phase 1,2-dihydro-3H-1,4-benzodiazepin-2-ones can possess rather different conformations, and in some cases the solution can contain 2 conformers. It seemed rather important to us to answer the question if the pharmacological activity of the studied benzodiazepines and their affinity for BDR can vary depending on their conformation and ratio of the conformer.

According to ERC mass spectrometry the more active compounds of the series XXXX have preferred conformation C_1 (pseudoboat), while less active member of this series have the preferred conformation C_2 (the flattened one with the anticlinal conformation of the amide group).

On this basis we have suggested that the conformation C_1 is more active than the conformation C_2, for the interaction of 1,2-dihydro-3H-1,4-benzodiazepin-2-ones with BDR.

To confirm this suggestion one can consider the fact that the affinity for BDR is greater for those compounds in solution which have higher cyclic dimers (Fig. 5)

Since the dimers exist only as C_1 conformers, it is suggested that the increase of the content of conformers C_1 results in increasing affinity for 1,2-dihydro-3H-1,4-benzodiazepin-2-ones towards BDR [15].

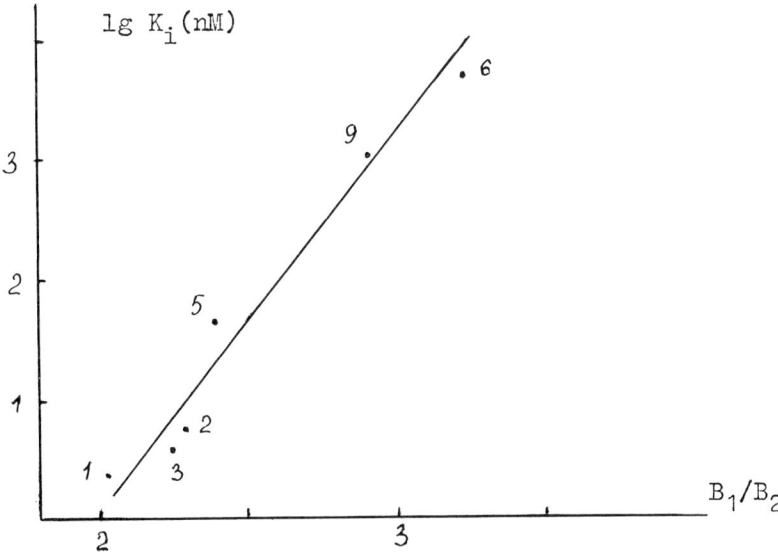

Fig. 5. Relationship between the reciprocal of affinity (K_i) ant the ratio of the integral intensities of absorption bands of NH groups of monomeric (B_1) and cyclically dimerized (B_2) benzodiazepinones.

REFERENCES
1. A.V. Bogatsky, S.A. Andronati, N.Ya. Golovenko, "Tranquilizers. 1,4-Benzodiazepines and the Related Structures", Naukova Dumka, Kiev, 1980.
2. G. Gilli, P.A. Borea, V. Bertolasi, M. Sagerdoti, "Molecular Structure and Biological Activity, Ed. by J.E. Griffin and W.L. Duak, Elsevier Sci. Publ. Co., New York, 1982.
3. L.H. Sternbach, "Benzodiazepines, Ed. by S. Garattinit et al., Raven Press, New York, 1973.
4. S.A. Andronati, A.S. Yavorsky, V.M. Chepelev et al., "Mechanisms of Action of Anxiolytic, Anticonvulsant and Hypnotic Agents", Naukova Dumka, Kiev, 1988.
5. R.A. Akhundov, "Molecular Basis of the Action of Psychotropic Agents", Moscow, 1986.
6. S.A. Andronati, T.A. Voronina, V.M. Chepelev, T.I. Korotenko, Khimiko-farmatsevticheskii Zhurnal, 1983, 1296.
7. S.A. Andronati and T.A. Voronina, Sov. Med. Rev. G. Neuropharm, 1990, 1, 77.
8. P. Linsheid, J. Lehn, Bull. Soc. Chim. France, 1967, 992.
9. A. Mannschreck, Chem. Ber., 1967, 100, 335.
10. V.I. Khvostenko, O.G. Khvostenko, G.S. Lomakin et al., Izvestiya AN SSSR, Ser. Khim., 1987, 1277.
11. S.A. Andronati, V.I. Khjvostenko, O.G. Khvostenko, et al., Doklady AN SSSR, 1989, 305, 99.
12. J. Bergman, A. Brynolf, Heterocycles, 1983, 20, 2145.
13. S.A. Andronati, I.A. Prokopenko, A.S. Yavorsky et al. Pharmazie, 1988, 43, 730.
14. P. Seiler, I. Zimmermann, Arzneim.-Forsch./Drug.Res., 1983, 33, 1519.
15. S.A. Andronati, I.A. Prokopenko, A.S. Yavorsky, et al., Zhurnal Organicheskoi Khimii, 1990, 60, 1151.

Design, Synthesis, and Evaluation of Functional Analogs of CC-1065

D. L. Boger

DEPARTMENT OF CHEMISTRY, PURDUE UNIVERSITY, WEST LAFAYETTE, INDIANA 47907, USA

1 INTRODUCTION

(+)-CC-1065 (**1**), a potent antitumor antibiotic isolated from cultures of *Streptomyces zelensis*,[1] has been shown to possess potent in vitro cytotoxic activity, broad spectrum antimicrobial activity, and confirmed in vivo antitumor activity.[2-4] In a series of extensive investigations the site and mechanism of the (+)-CC-1065 antitumor activity have been related to its sequence selective minor groove covalent alkylation within two, five base-pair consensus sequences [5´-d(A/GNTTA)-3´ and 5´-d(AAAAA)-3´] that has been demonstrated to proceed by 3´ adenine N-3 alkylation of the electrophilic cyclopropane present in the left-hand subunit.[2-11] Herein, we summarize studies designed to address the structural and functional origin of the affinity and selectivity of the CC-1065 interaction with duplex DNA, a process we have come to refer to as "accessible hydrophobic binding driven bonding."[12] Precisely what process this is intended to represent and the test of this understanding through the design, synthesis, and evaluation of three functional analogs of (+)-CC-1065 are detailed, Figure 1. The first agent examined which we have come to refer to as CPI-CDPI$_2$ constitutes a precise functional analog of CC-1065 that was anticipated and found to embody the precise structural features of the natural product responsible for its sequence selective covalent alkylation of DNA and the resultant biological properties. The second class of agents examined that we have come to refer to as CI-CDPI$_x$ constitute agents possessing the required minimum (but not necessarily optimal) potent pharmacophore of the alkylation subunit of CC-1065 and their evaluation provided far more insight into the relative importance of the alkylation subunit of the agents than anticipated. Given the information derived from the examination of these initial two classes of agents, the design and evaluation of a third class of analogs, CBI-CDPI$_x$, are detailed. The CBI-based agents have been found to constitute enhanced functional analogs of CC-1065 that possess greater in vitro cytotoxic potency, enhanced stability, and increased productive DNA binding properties.

From these studies, if there is a single important feature of the DNA:CC-1065 interaction that we would like to highlight, it is the important role that the agent noncovalent binding selectivity may play in the selectivity of the covalent alkylation of DNA and its potential origin. We will summarize the experimental studies that have led to these conclusions and which serve to further refine conclusions drawn from early studies.[2-11]

Figure 1

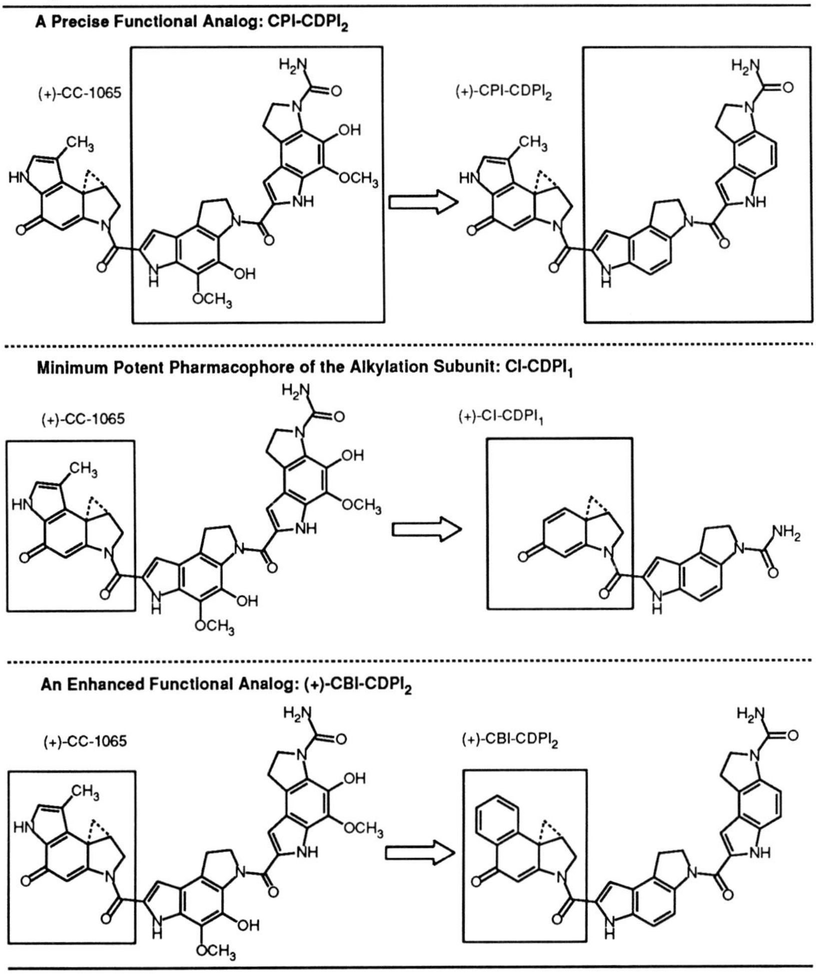

A Precise Functional Analog: CPI-CDPI$_2$

(+)-CC-1065   (+)-CPI-CDPI$_2$

Minimum Potent Pharmacophore of the Alkylation Subunit: CI-CDPI$_1$

(+)-CC-1065   (+)-CI-CDPI$_1$

An Enhanced Functional Analog: (+)-CBI-CDPI$_2$

(+)-CC-1065   (+)-CBI-CDPI$_2$

2 BACKGROUND

Since the initial disclosure of the structure of (+)-CC-1065[1] and its confirmation in a single crystal X-ray structure determination,[11] the agent and structural analogs have been the subject of extensive synthetic[12-18] and biological studies that have been extensively reviewed.[2-4,12-13] In particular, the synthetic efforts of the Upjohn group, Magnus, Rees, Cava, Sundberg, and Umezawa complement the studies of our own that are detailed herein.[12] Throughout the time that our initial synthetic efforts were in progress and directed at achieving a total synthesis of (+)-CC-1065 and ent-(-)-CC-1065,[19-23] the Upjohn group in collaboration with Hurley and coworkers established the fundamental features characterizing the DNA:CC-1065 interaction. In these studies, they established the nature of the (+)-CC-1065 irreversible covalent alkylation of double-stranded DNA that was confirmed through isolation and characterization of the adenine N-3 linked CC-1065 covalent adduct thermally released from DNA, Figure 2.[2-4,7] This and the clear

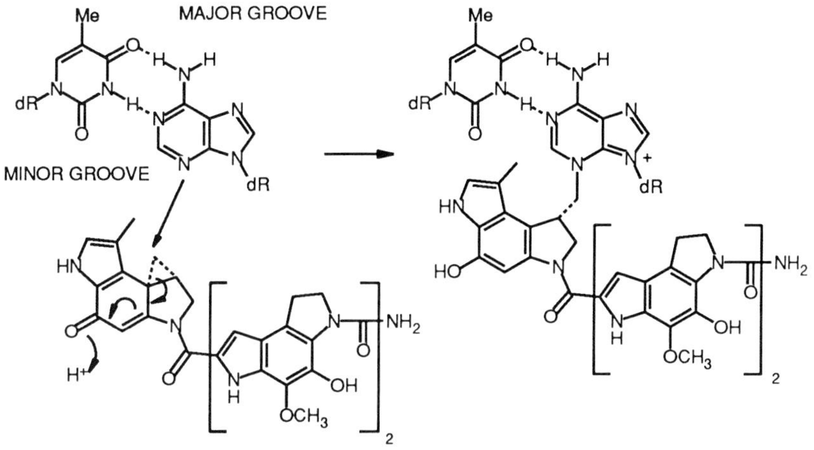

Functional Analogs of CC-1065 105

Figure 2 Adenine N-3 Alkylation of CC-1065
3' Adenine of 5'-d(AAAAA)-3' and 5'-d(G/ANTTA)-3'

identification of the sites of covalent alkylation and the agent binding directionality that may be derived from autofootprinting and classical footprinting studies on end-labeled restriction fragments of duplex DNA led to the identification of preferred or accessible sequences.[2-4,9-10] With this information in hand, we initiated a program to study the nature of the DNA:CC-1065 interaction with the intent of clearly defining the origin of the agents affinity and alkylation selectivity for double-stranded DNA.

In the course of the early investigations, Hurley and the Upjohn group first detailed the lack of a detectable interaction of CPI, the alkylation subunit of CC-1065, with duplex DNA[6] and consequently CC-1065 initially was represented as a reactive electrophile (CPI) linked to a DNA binding agent (PDE-I$_2$). The more recent demonstration by the Upjohn and Hurley groups that (+)-N-acetyl-CPI, a simple derivative of the (+)-CC-1065 alkylation subunit, exhibits a comparable albeit substantially less intense (ca. 10000x) covalent alkylation of duplex DNA has led to the revised conviction that the left-hand subunit of CC-1065 plays the dominant role in controlling the properties of the agents including its DNA covalent alkylation sequence selectivity.[4,10] The description of this work appeared at approximately the same time that our studies revealed that central and right-hand subunits of CC-1065 were contributing to its DNA binding affinity and selectivity in a fashion that exceeded the prescribed role of simply providing DNA affinity properties to a selective alkylation agent (CPI). Much of the impetus for our continued investigation was derived from the observation that the individual subunits of CC-1065, CPI[6] and PDE-I$_2$ methyl ester,[12,29] possessed comparable albeit modest in vitro cytotoxic activity and a resulting interest in defining the synergism of biological activity that accompanies their coupling to provide CC-1065, Figure 3.

3 NONCOVALENT BINDING AFFINITY AND SELECTIVITY

In initial efforts to identify the structural features of CC-1065 that may contribute to its high affinity, noncovalent minor groove binding affinity or selectivity, an extensive molecular modeling study revealed

Figure 3

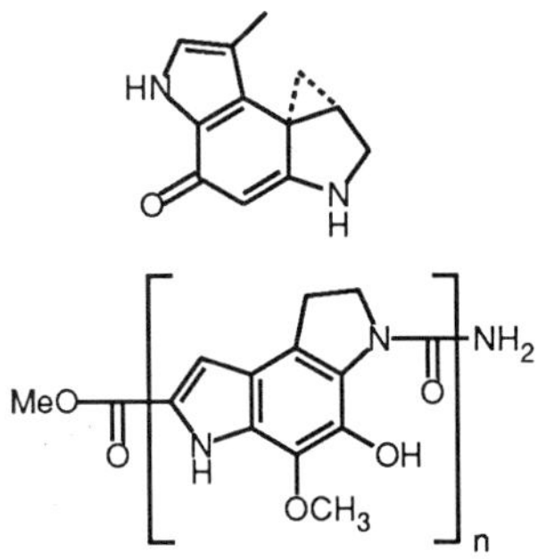

	IC_{50} (10^{-6} g/mL) L1210
CPI	0.2
PDE-I_1	> 2.0
PDE-I_2	0.04
CC-1065	0.00001

testable insights into the nature and potential origin of the noncovalent interaction. Without providing the details of this study, some of which have been described,[12] the results suggested: (1) that oligomers of CDPI (CDPI$_n$) may possess DNA binding affinity and specificity comparable to that of oligomers of PDE-I (PDE-I$_n$), (2) that the DNA:agent affinity may be attributed predominantly to stabilizing van der Waals contacts involving stabilizing interactions principally between the DNA deoxyribose residues and the agent, (3) that the stabilizing interaction could be achieved without significantly distorting the standard B-DNA geometry, (4) that the agent DNA-bound conformation closely follows that of the agents accessible low energy, helical conformation inherent in the CDPI$_n$ structure, and (5) that a significant preference for A-T versus G-C rich noncovalent minor groove binding selectivity may be anticipated. In addition, and in contrast to our intuition, the studies suggested no apparent noncovalent binding selectivity that could be attributed to the presence of the CC-1065 methoxy and hydroxy substituents and only a small role for their contribution to DNA noncovalent binding affinity.

Consequently, the oligomers of CDPI, CDPI$_1$ - CDPI$_5$, and PDE-I, PDE-I$_1$ - PDE-I$_3$, were prepared[21,25-26] and their DNA binding properties examined.[24] The comparative DNA binding properties of CDPI$_n$ methyl esters (n = 1-5) versus PDE-I$_n$ methyl esters (n = 1-3) provided experimental evidence for the intrinsic importance of stabilizing hydrophobic binding and noncovalent van der Waals contacts dominant in the CC-1065:B-DNA minor groove binding. High affinity minor groove binding to DNA was established through: (1) the observation of CDPI$_3$ binding (UV) but not unwinding of supercoiled DNA (ϕ174 RFI DNA) thus excluding intercalative binding, (2) the observation of CDPI$_3$ binding to T$_4$ phage DNA (UV, ΔT_m) in which the major groove is occluded by glycosylation thus excluding major groove binding, (3) the observation of salt (Na$^+$) concentration independent high affinity CDPI$_3$ binding to poly[dA]·poly[dT] thus excluding simple electrostatic binding to the DNA phosphate backbone, and further inferred through (4) the observation of an intense induced circular dichroism (ICD, poly[dA]·poly[dT] and poly[dG]·poly[dC] $[\phi]^{23}_{358}$ = 24,000 and 23,500).

Figure 4

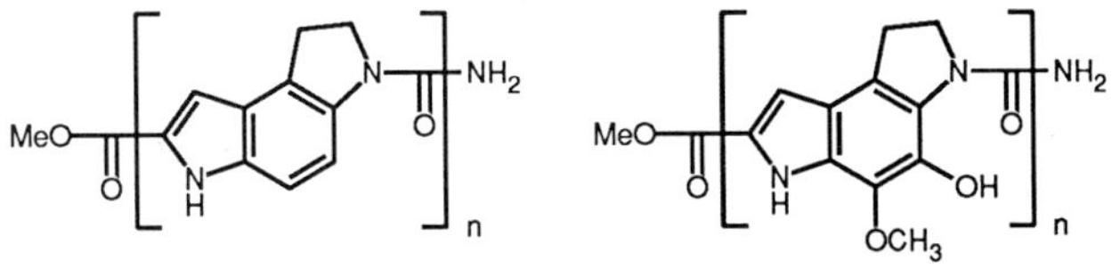

CDPI$_n$ n = 1 - 5

PDE-I$_n$ n = 1 - 3

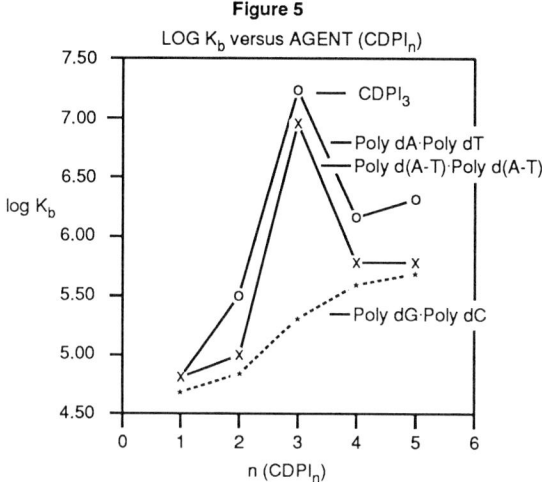

Figure 5
LOG K_b versus AGENT (CDPI$_n$)

CDPI$_3$	K_b	$\Delta G°$ (298° K)
Poly dA·Poly dT	1.8×10^7 M^{-1}	-9.9 kcal
Poly d(A-T)·Poly d(A-T)	8.9×10^6 M^{-1}	-9.5 kcal
Poly dG·Poly dC	2.1×10^5 M^{-1}	-7.3 kcal

	a Exp.	b Calcd.
Poly dA·Poly dT		
ln K_b (CDPI$_2$)/ln K_b (PDE-I$_2$)	0.77	0.87
ln K_b (CDPI$_2$)/ln K_b (CDPI$_3$)	0.66	0.79
ln K_b (PDE-I$_2$)/ln K_b (CDPI$_3$)	0.98	0.90
ln K_b (CDPI$_2$-bound CDPI$_4$)/ln K_b (CDPI$_3$)	0.84	0.87
ln K_b (CDPI$_2$)/ln K_b (CDPI$_2$-bound CDPI$_4$)	0.79	0.90
CDPI$_3$		
ln K_b Poly dG·Poly dC/ln K_b Poly dA·Poly dT	0.74	0.86

a Exp. free energy of binding ($\Delta G°$, 298° K)
b Calcd. gas phase enthalpy of binding ($\Delta H°$)

From relative and absolute binding constant determinations it was demonstrated that: (1) CDPI$_3$ represents the optimum binding unit (25°C) within the oligomeric CDPI$_n$ agents studied (CDPI$_3$ > CDPI$_5$ ≅ CDPI$_4$ ≥ CDPI$_2$ > CDPI$_1$), (2) CDPI$_3$ minor groove binding spans five base-pairs or one-half a helix turn of double-stranded DNA which represents the largest accessible binding site available for synchronous binding of the rigid agent termini (poly[dA]·poly[dT] and poly[d(A-T)]·poly[d(A-T)] agent:base-pair ratio = 0.2), (3) partial bound forms of the larger oligomer agents, i.e., CDPI$_2$-bound CDPI$_4$, constitute productive and relevant DNA:agent complexes, (4) the PDE-I$_n$:DNA complexes are thermodynamically more stable and kinetically more labile than the corresponding CDPI$_n$:DNA complexes, (5) the qualitative and relative quantitative free energies of binding of the agents to DNA follow trends depicted in a quantitative molecular modeling study from which estimates

Figure 6

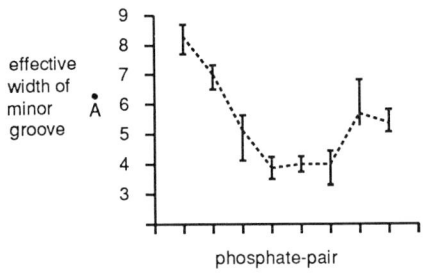

Range and average from four X-ray crystal structures

d(CGCGATATCGCG)$_2$:netropsin
Reference 27
d(CGCAAATTTGCG)$_2$:distamycin
Reference 28
d(CGCGAATTCGCG)$_2$
Reference 29
d(CGCA$_6$GCG):d(GCGT$_6$CGC)
Reference 30

of the gas phase enthalpies of binding were available (poly[dA]·poly[dT] and poly[d(A-T)]·poly[d(A-T)] > poly[dG]·poly[dC]: PDE-I$_3$ > CDPI$_3$ > PDE-I$_2$ > CDPI$_2$-bound CDPI$_4$ > CDPI$_2$ > CDPI$_1$), (6) the absence of CDPI$_n$ functionality that would permit stabilization of DNA:CDPI$_n$ complexes through hydrogen bonding or substantial stabilizing electrostatic interactions suggests that the binding affinity of the agents is derived principally if not exclusively from hydrophobic binding and stabilizing van der Waals contacts (affinity = hydrophobic binding), and (7) the CDPI$_n$ agents exhibit a substantial preference for A-T rich versus G-C rich DNA minor groove binding, Figure 5, which represents preferential binding to minor groove regions in which stabilizing agent:DNA van der Waals contacts would be enhanced with the deeper penetration of the agents into the narrower A-T rich DNA minor groove (specificity = accessible hydrophobic binding). Consequently, the studies suggest that (+)-CC-1065 is best represented as a selective alkylating group superimposed upon the rigid CDPI$_3$ skeleton rather than as reactive alkylating agent (CPI) attached to a DNA binding agent (PDE-I$_2$). The initial noncovalent DNA binding of the agents may be expected to exhibit a substantial preference for A-T rich DNA minor groove binding and is optimal with the properties and size of the rigid, hydrophobic trimer (CDPI$_3$/PDE-I$_3$). Among many features that have been suggested to contribute to an agents minor groove binding selectivity, two structural characteristics of an A-T base-pair and of a run of A-T base-pairs may prove to be of particular importance to agents that rely principally or exclusively on stabilizing van der Waals contacts and hydrophobic binding for noncovalent binding stabilization. First, a G-C base-pair possesses an amino substituent that protrudes into the minor groove and a comparable substituent is not present with an A-T base-pair, Figure 6. This sterically more accessible minor groove region adjacent to an A-T base-pair permits the deeper minor groove penetration of an agent adjacent to an A-T versus G-C base-pair and enhances the binding stabilization derived from van der Waals contacts. In addition, recent X-ray crystallographic studies of deoxyoligonucleotides that contain central to their structure a run of A-T base-pairs have revealed

Figure 7

A Precise Functional Analog: CPI-CDPI$_2$

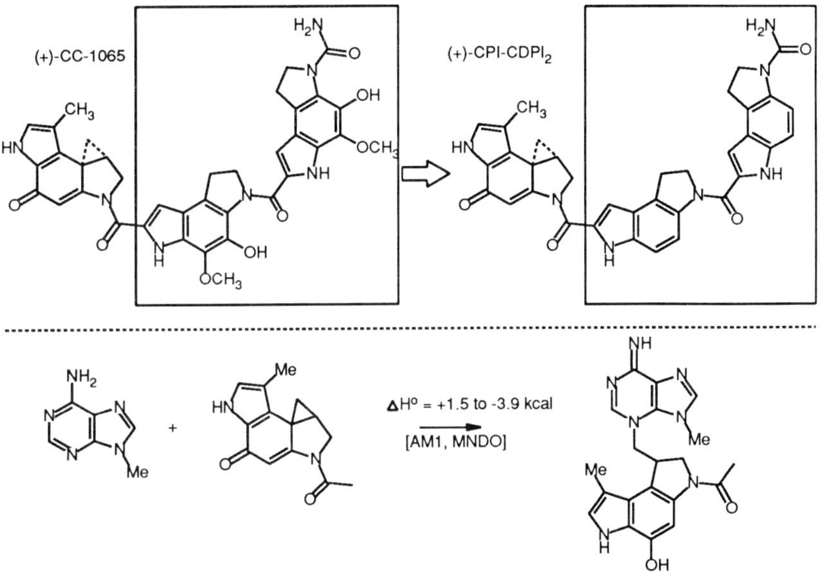

characteristic conformational features notable of which is the constricted width of the A-T rich minor groove,[27-30] Figure 6. Illustrated in Figure 6 is a plot of the effective width of the minor groove (P - P distance - 5.8 Å) versus the phosphate-pair for four X-ray crystal structures of the deoxyoligonucleotides as detailed by Wang and Rich.[27] This important conformational feature of double-stranded DNA possessing runs of A-T base-pairs may contribute prominently to the noncovalent binding selectivity of the agents detailed herein which depend on stabilizing van der Waals contacts for noncovalent binding stabilization. The agents initial noncovalent complexes with DNA may be expected to form preferentially within the narrower, sterically more accessible A-T rich minor groove where the stabilizing van der Waals contacts are most fully achieved: accessible hydrophobic binding.

4 CPI-CDPI$_2$: A PRECISE FUNCTIONAL ANALOG OF CC-1065

The comparable noncovalent DNA binding behavior of CDPI$_n$ methyl esters and PDE-I$_n$ methyl esters suggested that the incorporation of CDPI$_2$ into the CC-1065 structure in place of PDE-I$_2$ could potentially provide an agent with comparable or indistinguishable properties.[12,19] This agent, CPI-CDPI$_2$, seemed particularly important to examine given the growing formulation that the polar functionality of (+)-CC-1065 was contributing in a significant fashion[31] to the properties of the agent, Figure 7.

Prior to detailing the preparation of this agent, one additional fundamental feature of the CC-1065 covalent alkylation of DNA merits further comment. The activated cyclopropane of the alkylation subunit of CC-1065 constitutes a modest electrophile as evidenced by its chemical stability. Simple derivatives of CPI, i.e. N-BOC-CPI, are stable in aqueous or protic solutions at pH 5-7, undergo acid-catalyzed solvolysis with opening of the cyclopropane only slowly ($t_{1/2}$ = 37 h, pH = 3), and typically suffer N-deacylation versus cyclopropane ring opening in the presence of nucleophiles under basic conditions. Representative of this

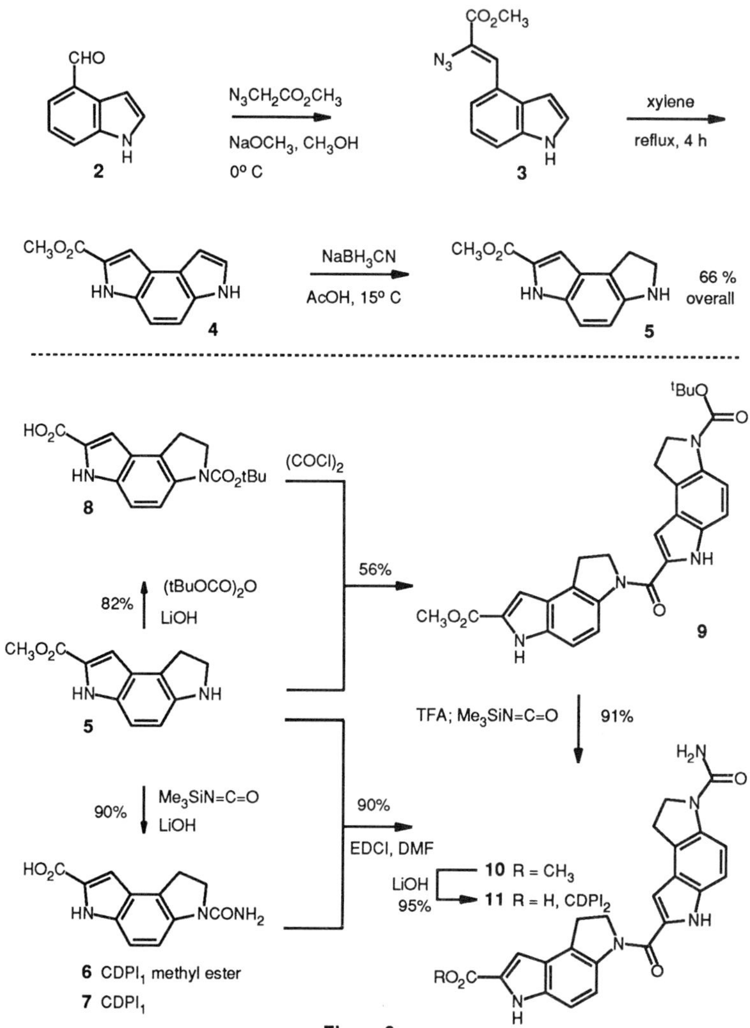

Figure 8

modest electrophilic character of CPI is the thermal neutral nature of adenine N-3 covalent alkylation of N-acetyl-CPI, Figure 7. At present, the adenine N-3 alkylation of derivatives of CPI has been achieved only with intact double-stranded DNA, has been suggested to represent a DNA (Lewis) acid-catalyzed reaction,[4] and has not been observed with free adenine. To us, this has suggested that the dominant noncovalent binding of the agents serves to stabilize (drive) the thermal neutral adenine N-3 covalent alkylation reaction: accessible hydrophobic binding-driven-bonding.

$CDPI_2$ for incorporation into $CPI-CDPI_2$ was prepared in six steps (50-55% overall) as detailed in Figure 8.[25] A precursor to the CPI subunit of the agent was prepared as detailed in Figure 9,[19,40] resolved through chromatographic separation of the diastereomers 24, Figure 10,[32] and incorporated into the synthesis of (±)-, (+)-, and (-)-$CPI-CDPI_2$ as detailed in Figure 11.[19,32]

Functional Analogs of CC-1065

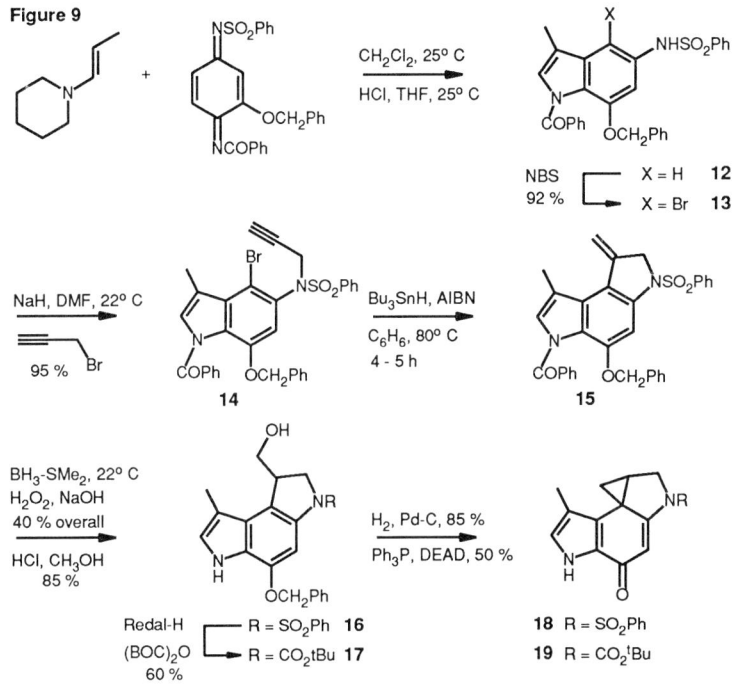

Figure 9

Figure 10

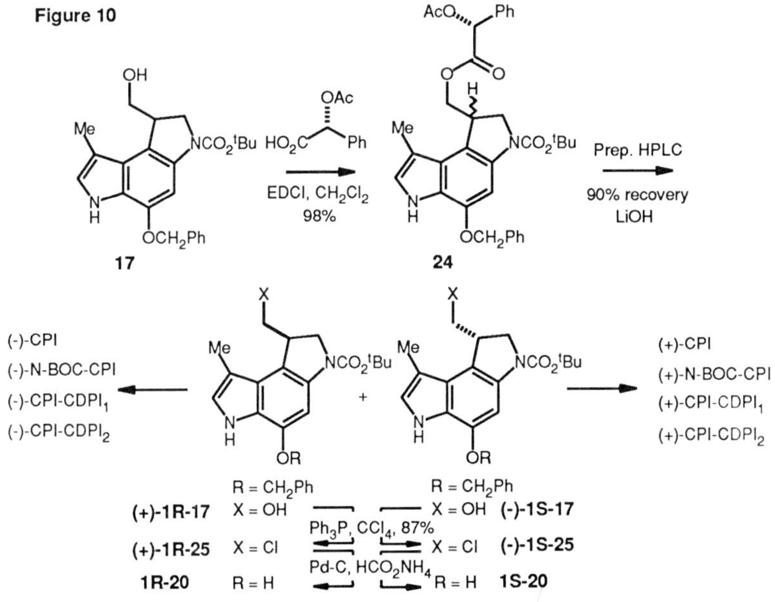

Figure 11

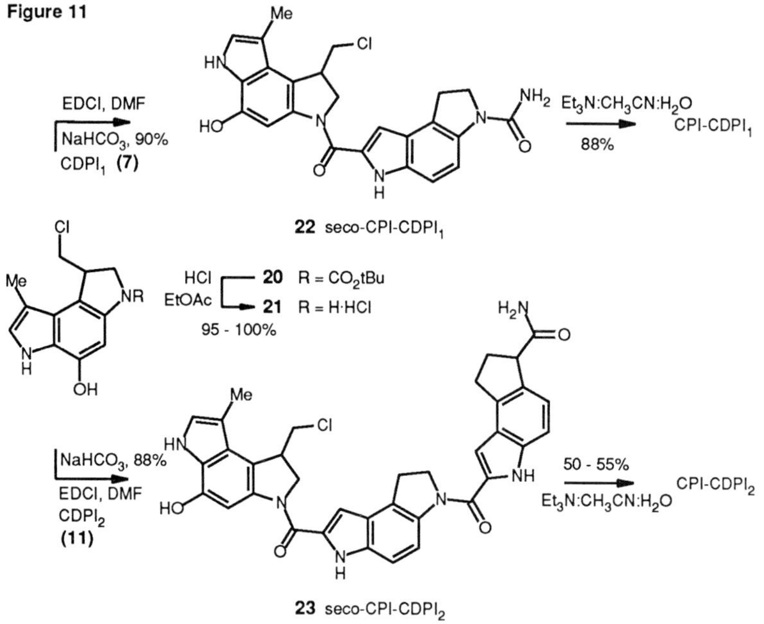

22 seco-CPI-CDPI₁

20 R = CO₂tBu
21 R = H·HCl

23 seco-CPI-CDPI₂

Initial studies comparing the properties of CC-1065 and CPI-CDPI₂ rested with their in vitro cytotoxic evaluation which revealed that the agents were not distinguishable, Figure 12. Both enantiomers of CPI-CDPI₂ displayed potent and indistinguishable in vitro cytotoxic activity that proved to be equivalent to that observed with (+)-CC-1065/ent-(-)-CC-1065. Most satisfying beyond this indistinguishable level of cytotoxic potency of CPI-CDPI₂/CC-1065 was the observation of the equipotent activity of the unnatural enantiomers. This unique property of CC-1065 is not embodied in simplified analogs bearing the CPI alkylation subunit[4,6,12] but is inherent in CPI-CDPI₂. Similar observations have been made in the Upjohn efforts.[31]

A more definitive demonstration that CPI-CDPI₂ embodies the precise structural and functional features of CC-1065 responsible for its sequence selective covalent alkylation of DNA and the resulting biological activity, was derived from the examination of their DNA binding properties.[33] The profile (selectivity) of the (+)-CC-1065 and (+)-CPI-CDPI₂ covalent alkylation of DNA has proven indistinguishable, Figure 13. This indistinguishable alkylation profile proved particularly revealing when compared to the alkylation profile of the series of structurally related agents detailed in the next section and which was facilitated through development of useful protocol for securing substantial quantities of singly ³²P-5´-end-labeled double-stranded DNA cloned into the SmaI site of M13mp10 suitable for the conduct of such studies.[34-35]

A model of the (+)-CC-1065 and (+)-CPI-CDPI₂ binding to the high affinity site of w794 double-stranded DNA is illustrated in Figure 14. Significant to the support of the potential origin the A-T rich noncovalent binding selectivity of the agents and its contribution to the DNA alkylation selectivity was the observation that the model complex

Figure 12 Comparative In Vitro Cytotoxic Activity

Agent	Configuration	L1210 ID$_{50}$ (10^{-11} g/mL)	
		CC-1065	CPI-CDPI$_2$
(natural)		1.1	
(synthetic)	natural	1.2	1.3
(synthetic)	enantiomeric	1.3	1.3

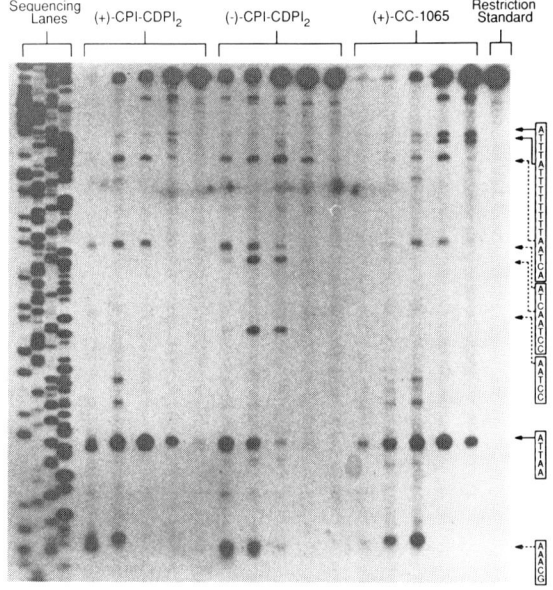

Figure 13. Thermally-induced strand cleavage of a double-stranded SV40 fragment (144 b.p.; nucleotide # 138-5238, clone w794) after 24 h incubation of agent:DNA at 4°C followed by removal of unbound agent and 30 min incubation at 100°C; 8% denaturing poly(acrylamide) gel electrophoresis, 1100V/6 h and autoradiography. Lane 1, Sanger G reaction; Lane 2, Sanger C reaction; Lane 3, Sanger A reaction; Lane 4, Sanger T reaction; Lanes 5-9, (+)-CPI-CDPI$_2$ (5.6 x 10^{-4} - 5.6 x 10^{-8} M); Lanes 10-14, (-)-CPI-CDPI$_2$ (5.6 x 10^{-4} - 5.6 x 10^{-8} M); Lanes 15-19, (+)-CC-1065 (5.6 x 10^{-4} - 5.6 x 10^{-8} M); Lane 20, DNA standard.

Figure 14. Model of 5′-d(CTCAATTAGTC)-3′:(+)-CC-1065 and 5′-d(CTCAATTAGTC)-3′:(+)-CPI-CDPI$_2$ constituting the high affinity alkylation site within w794 DNA. DNA and DNA:agent complexes were generated with MacroModel (Version 2.5) and subjected to full structure energy optimization (RMS = 0.3 kJ/Å, AMBER force field supplemented with parameters for the agents).

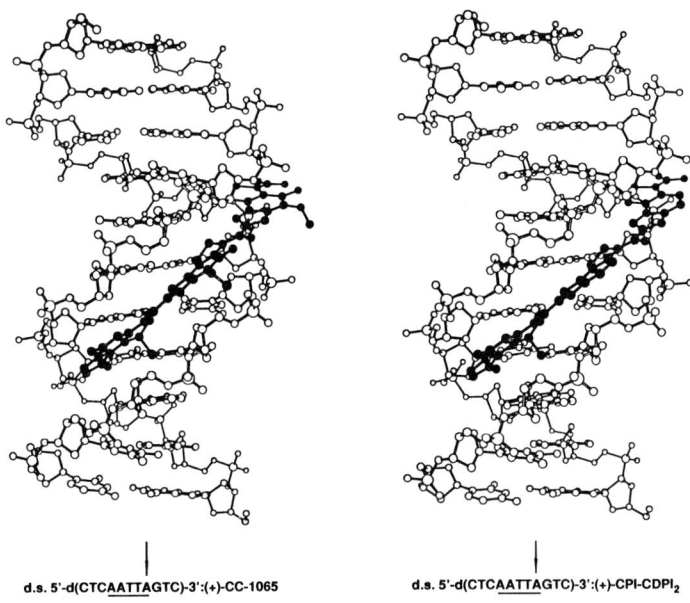

d.s. 5′-d(CTC<u>AATT</u>AGTC)-3′:(+)-CC-1065   d.s. 5′-d(CTC<u>AATT</u>AGTC)-3′:(+)-CPI-CDPI$_2$

exhibits the characteristic constriction of the width of the minor groove through the A-T rich region analogous to that observed in several X-ray crystal structures, Figure 15.

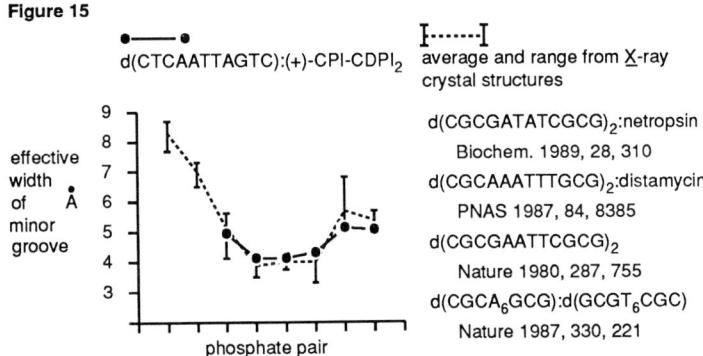

Figure 15

●——● d(CTCAATTAGTC):(+)-CPI-CDPI$_2$

I······I average and range from X-ray crystal structures

d(CGCGATATCGCG)$_2$:netropsin
Biochem. 1989, 28, 310

d(CGCAAATTTGCG)$_2$:distamycin
PNAS 1987, 84, 8385

d(CGCGAATTCGCG)$_2$
Nature 1980, 287, 755

d(CGCA$_6$GCG):d(GCGT$_6$CGC)
Nature 1987, 330, 221

effective width of minor groove (Å)

phosphate pair

Figure 16

Aborted and Extended Functional Analogs: CPI-PDE-I$_n$ and CPI-CDPI$_n$

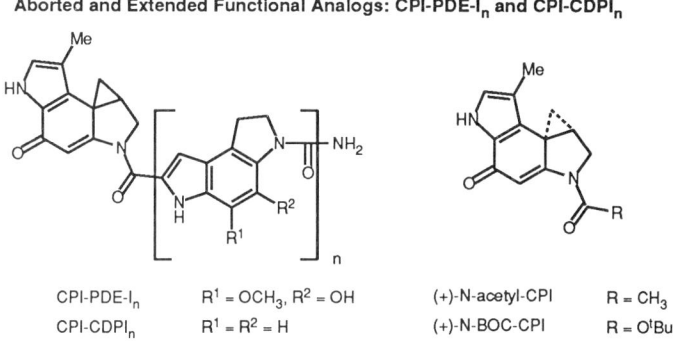

CPI-PDE-I$_n$	R^1 = OCH$_3$, R^2 = OH	(+)-N-acetyl-CPI	R = CH$_3$
CPI-CDPI$_n$	R^1 = R^2 = H	(+)-N-BOC-CPI	R = O^{t}Bu

5 ABORTED AND EXTENDED FUNCTIONAL ANALOGS OF CC-1065: CPI-PDE-I$_n$ AND CPI-CDPI$_n$

In efforts to establish the functional role of each of the subunits of CC-1065, (+)-N-acetyl- and (+)-N-BOC-CPI, simple derivatives of the alkylation subunit, and the aborted and extended agents CPI-PDE-I$_1$, CPI-CDPI$_1$, and CPI-CDPI$_3$ were prepared,[33] Figure 16. The preparation was accomplished following the protocols detailed in Figure 11 for CPI-CDPI$_1$. Although a number of interesting observations were derived from this study,[33] only a few of the most pertinent are discussed herein. First, (+)-CPI-PDE-I$_1$ displayed in vitro cytotoxic activity at levels indistinguishable from (+)-CC-1065 and its unnatural enantiomer, (-)-CPI-PDE-I$_1$, proved inactive within the limits of detection ($\leq$ 1% natural enantiomer contaminant), Figure 17. Similar observations were made with (+)- and (-)-CPI-CDPI$_1$ while both enantiomers of CPI-CDPI$_3$ proved to be potent cytotoxic agents. As previously detailed, (+)-N-BOC-CPI proved to be approximately 10000x less potent. As detailed in the efforts of the Upjohn groups,[4-6] the trends in cytotoxic potency correlated with the agents relative intensity of the covalent alkylation of double-stranded DNA as estimated through an autofootprinting procedure. However, in contrast to the suggestion that a potential relationship between the agents cytotoxic potency and DNA binding properties lies with their relative rates of covalent alkylation,[4] the comparable potency of (+)-CC-1065, (+)-CPI-PDE-I$_1$, (+)-CPI-CDPI$_2$, (+)-CPI-CDPI$_1$, and (+)-CPI-CDPI$_3$ suggests that the relationship may reside in the simple event (not extent) of noncovalent binding stabilization of the thermal neutral covalent alkylation. As such, an additional functional role the agents central and/or right-hand subunits may be the restriction of the number of accessible adenine alkylation sites through preferential binding within the narrower, sterically more accessible A-T rich minor groove. The strict A-T preference for the first three base-pairs of the CC-1065 alkylation sequences may then reflect the combination of the initial 3'-adenine alkylation site coupled with the required noncovalent stabilization provided by the agents central subunit and its two base-pair A-T binding selectivity.[24,33]

Pertinent to our interests in defining the origin of the (+)-CC-1065 covalent alkylation selectivity, (+)-CPI-CDPI$_1$ and (+)-CPI-CDPI$_3$ were shown to possess profiles of DNA covalent alkylation that were subtly distinguishable from that of (+)-CC-1065/(+)-CPI-CDPI$_2$, Figure 18. More importantly, in addition to the reduced intensity of DNA covalent alkylation exhibited by N-BOC-CPI and N-acetyl-CPI (ca. 10000x), the DNA covalent alkylation profile of (+)-N-BOC-CPI/(+)-N-acetyl-CPI under our examination proved substantially different and less selective than that

Figure 17 Comparative In Vitro Cytotoxic Activity

Agent	Configuration	ID$_{50}$ (L1210)		Autofootprinting
		(10^{-11} g/mL)	(nM, rel ID$_{50}$)	
(+)-N-BOC-CPI	natural	10000	330, 0.00006	see text
(+)-CPI-PDE-I$_1$	natural	0.8	0.02, 1	+
(−)-CPI-PDE-I$_1$	enantiomeric	>125	> 2.4, < 0.01	+/−
(+)-CPI-CDPI$_1$	natural	1.7	0.04, 0.5	+
(−)-CPI-CDPI$_1$	enantiomeric	>270	> 6.3, < 0.003	+/−
(+)-CPI-CDPI$_3$	natural	1.6	0.02, 1	+
(−)-CPI-CDPI$_3$	enantiomeric	10	0.1, 0.2	+
(+)-CC-1065	natural	1.2	0.02, 1	+
(−)-CC-1065	enantiomeric	1.3	0.02, 1	+
(+)-CPI-CDPI$_2$	natural	1.3	0.02, 1	+
(−)-CPI-CDPI$_2$	enantiomeric	1.3	0.02, 1	+

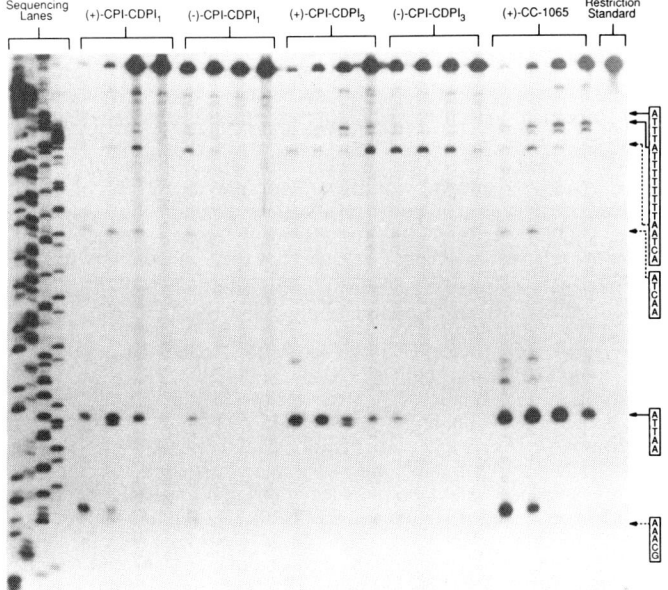

Figure 18. Thermally-induced strand cleavage of a double-stranded SV40 fragment (144 b.p.; nucleotide # 138-5238, clone w794) after 24 h incubation of agent:DNA at 4°C followed by removal of unbound agent and 30 min incubation at 100°C; 8% denaturing poly(acrylamide) gel, 1100V/6 h and autoradiography. Lane 1, Sanger G reaction; Lane 2, Sanger C reaction; Lane 3, Sanger A reaction; Lane 4, Sanger T reaction; Lanes 5-8, (+)-CPI-CDPI$_1$ (5.6 x 10^{-5} - 5.6 x 10^{-8} M); Lanes 9-12, (−)-CPI-CDPI$_1$ (5.6 x 10^{-5} - 5.6 x 10^{-8} M); Lanes 13-16, (+)-CPI-CDPI$_3$ (5.6 x 10^{-5} - 5.6 x 10^{-8} M); Lanes 17-20, (−)-CPI-CDPI$_3$ (5.6 x 10^{-5} - 5.6 x 10^{-8} M); Lanes 21-24, (+)-CC-1065 (5.6 x 10^{-5} - 5.6 x 10^{-8} M); Lane 25, control DNA.

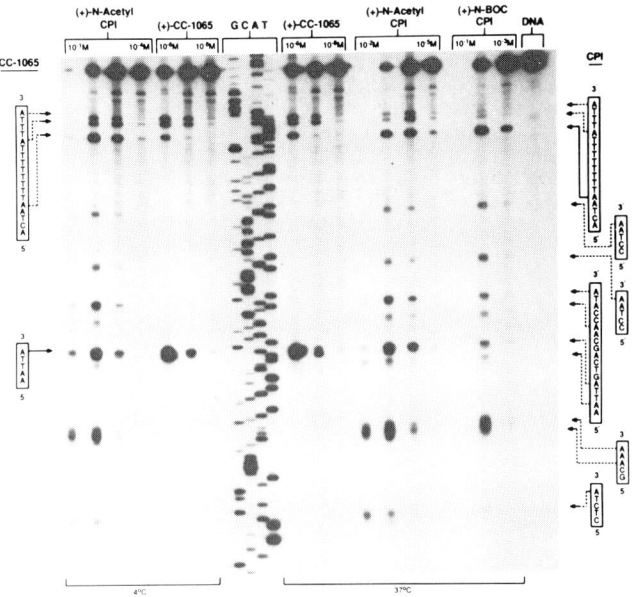

Figure 19. Thermally-induced strand cleavage of a double-stranded SV40 fragment (144 b.p.; nucleotide # 138-5238, clone w794) after 24 h incubation of agent:DNA at 4°C or 37°C followed by removal of unbound agent and 30 min incubation at 100°C; 8% denaturing poly(acrylamide) gel and autoradiography. Lanes 1-4, (+)-N-acetyl-CPI (4°C, 2.5×10^{-1} - 2.5×10^{-4} M); Lanes 5-7, (+)-CC-1065 (4°C, 1×10^{-6} - 1×10^{-8} M); Sanger G, C, A, and T reactions; Lanes 12-14, (+)-CC-1065 (37°C, 1×10^{-6} - 1×10^{-8} M); Lanes 15-18, (+)-N-acetyl-CPI (37°C, 2.5×10^{-2} - 2.5×10^{-5} M); Lanes 19-21, (+)-N-BOC-CPI (37°C, 2.5×10^{-1} - 2.5×10^{-3} M); Lane 22, control DNA.

of (+)-CC-1065, Figure 19. Consequently, contrary to the conclusions of others,[4,10] we interpreted this behavior to indicate that the sequence selectivity of the (+)-CC-1065 alkylation of DNA is derived only in part from the inherent selectivity of the CPI covalent alkylation. The A-T rich DNA noncovalent binding selectivity of the agents additionally restricts the number of accessible alkylation sites and productively contributes to the DNA covalent alkylation selectivity.[33] Later herein we detail comparisons that further suggest a more dominant role for an agents noncovalent binding selectivity in controlling the DNA covalent alkylation selectivity.

6 MINIMUM POTENT PHARMACOPHORE OF THE ALKYLATION SUBUNIT: CI-CDPI$_n$

Given the distinction in the relative importance attributed to the alkylation subunit of (+)-CC-1065 in contributing to its DNA covalent alkylation profile that have surfaced in our efforts and those of the

Hurley and the Upjohn groups, we elected to prepare a range of agents whose evaluation could potentially address the question directly. The first such agent considered incorporated the parent cyclopropylindole of the CC-1065 alkylation subunit, Figure 20. Consistent with intuition, computational studies suggested that the CI electrophile should be much more reactive than the authentic CPI subunit of CC-1065. As such, the agent could be anticipated to display an inherently less selective covalent alkylation of DNA. Thus, the comparisons of the DNA covalent alkylation profiles of N-BOC-CI/N-BOC-CPI and CI-CDPI$_2$/CC-1065/CPI-CDPI$_2$ were anticipated to provide direct evidence for the relative importance of the contributions of electrophile selectivity versus noncovalent binding selectivity to the CC-1065 covalent alkylation selectivity.

Complementary to our efforts on the preparation of the CPI subunit of CC-1065 that relied on a key 5-exo-dig aryl radical-alkyne free radical cyclization, the CI subunit construction employed a self-terminating 5-exo-trig aryl radical-alkene free radical cyclization in a concise preparation of **35-36**, Figure 21.[36-37]

The CI agents proved substantially more reactive than the analogous CPI agents and their successful preparation is detailed in Figures 22-23 and required attentive and respectful treatment to ensure their isolation and characterization.[36] Resolution of precursor **37** through the chromatographic separation of the diastereomers **45** provided the optically active agents (+)- and (-)-N-BOC-CI, (+)- and (-)-CI-CDPI$_1$, and (+)- and (-)-CI-CDPI$_2$, Figure 24.[37]

Consistent with expectations, the CI-based agents proved substantially more reactive than the CPI-based agents and this is reflected in the acid-catalyzed solvolysis rates of N-BOC-CI ($t_{1/2}$ = 35 sec, pH = 3; $t_{1/2}$ = 5.2 h, pH = 7) versus N-BOC-CPI ($t_{1/2}$ = 37 h, pH = 3; stable, pH = 7). Consequently, we were quite pleased with the in vitro cytotoxic activity of the agents in which (+)-CI-CDPI$_2$ proved to be only 500x less potent than (+)-CC-1065 and (+)-N-BOC-CI proved to be approximately 40-100x less potent than (+)-N-BOC-CPI despite the limited stability of the agents, Figure 25. As anticipated, the relative cytotoxic properties of the agents proved to correlate roughly with the relative intensity of the agents covalent alkylation of DNA as established through an autofootprinting protocol.

More important to our interests were the results of the comparative examination of the profiles of the CI-based agents double-stranded DNA covalent alkylation, Figures 26-27. The direct comparison of the DNA covalent alkylation profile of (+)-, (-)- and (±)-N-BOC-CI versus that of (+)-N-BOC-CPI/(+)-N-acetyl-CPI within w794 double-stranded DNA revealed a striking similarity between even racemic (±)-N-BOC-CI and the optically active CPI derivatives although the enhanced reactivity of N-BOC-CI results in a diminished selectivity among the alkylation sites common to both classes of agents. This striking similarity in the DNA covalent alkylation profiles provides the experimental evidence to support of the assignment of CI as the potent pharmacophore of the alkylation subunit (CPI) of CC-1065. Moreover, the sites of covalent alkylation available to N-BOC-CI proved independent of the absolute configuration of the electrophile and suggests a nondiscriminate alkylation event.

In sharp contrast, (+)-CI-CDPI$_2$ exhibits a profile of DNA covalent alkylation that is strikingly similar to that of (+)-CC-1065, Figure 27. In the case of (+)-CI-CDPI$_2$, there is no characteristic intrinsic

Functional Analogs of CC-1065

Figure 20

Minimum Potent Pharmacophore of the Alkylation Subunit: Cl-CDPI$_n$

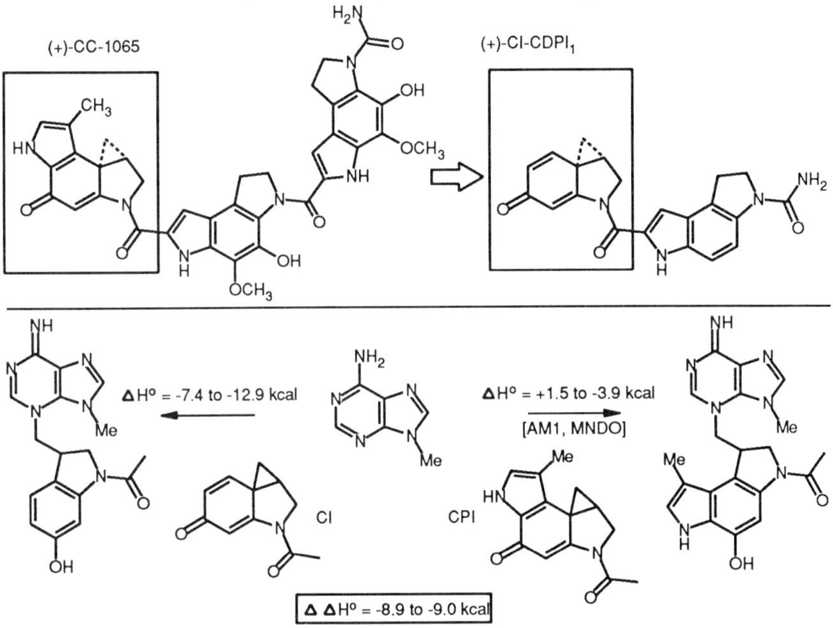

Figure 21

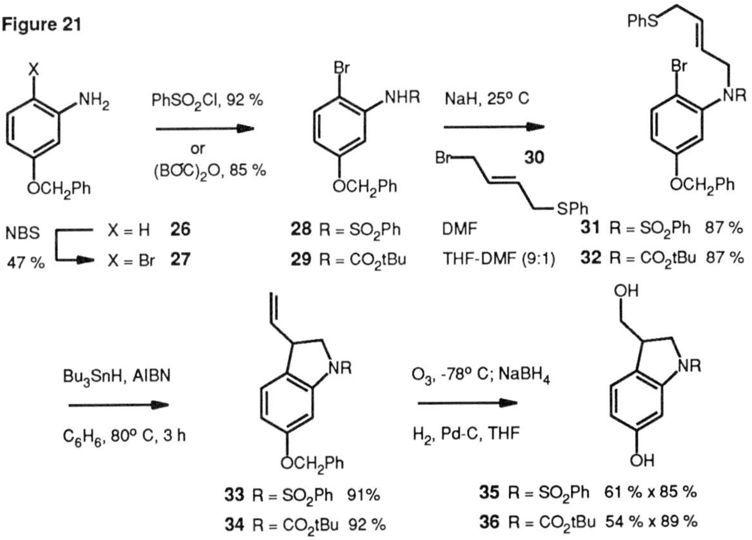

Figure 22

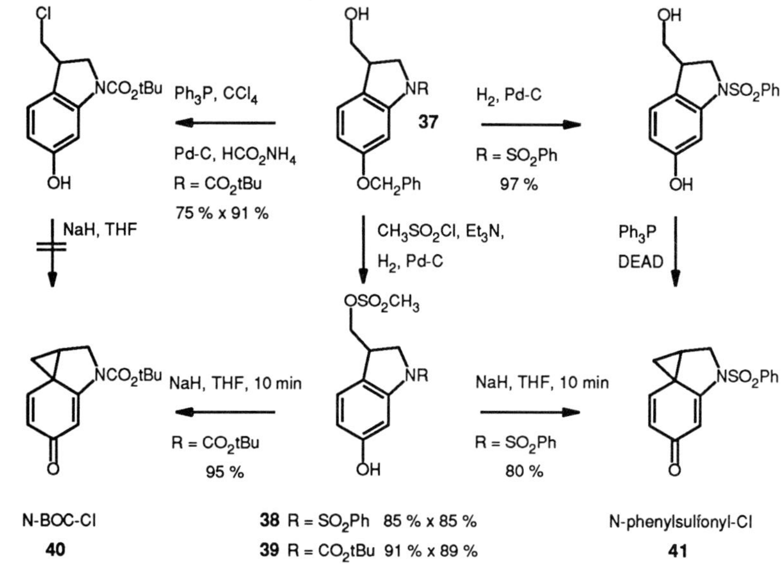

Figure 23

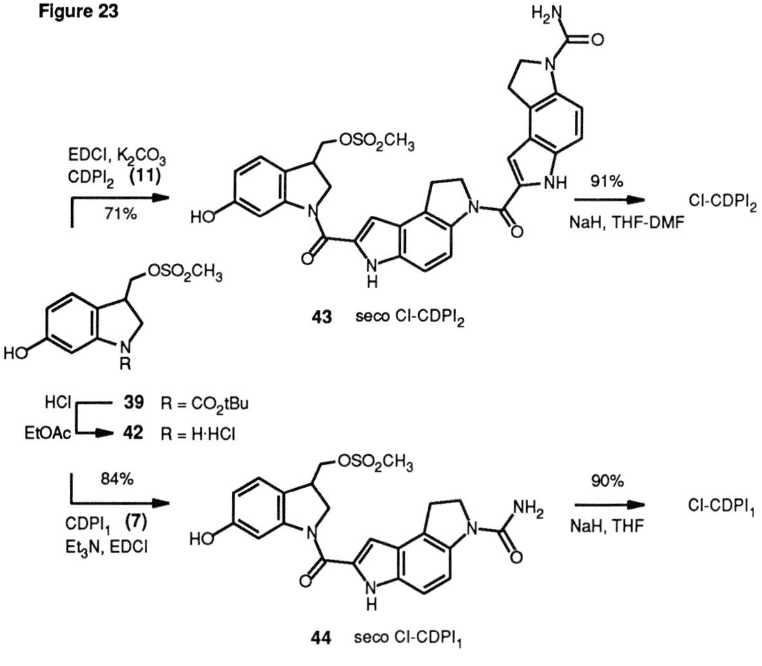

Figure 24

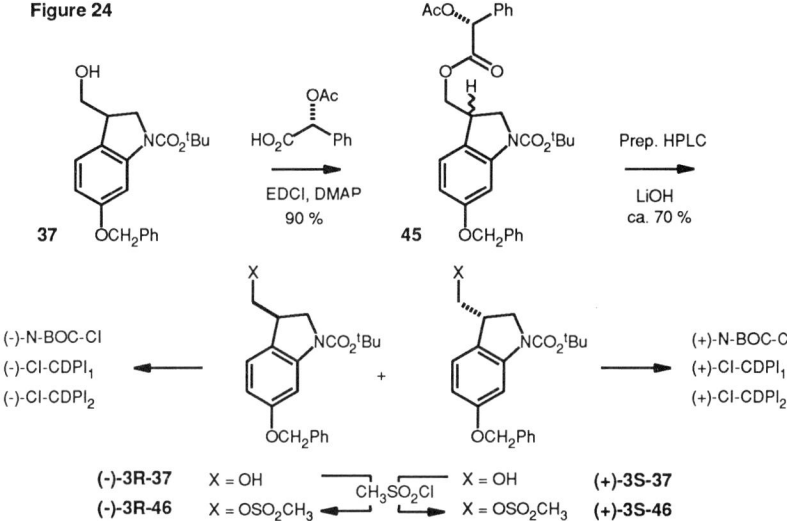

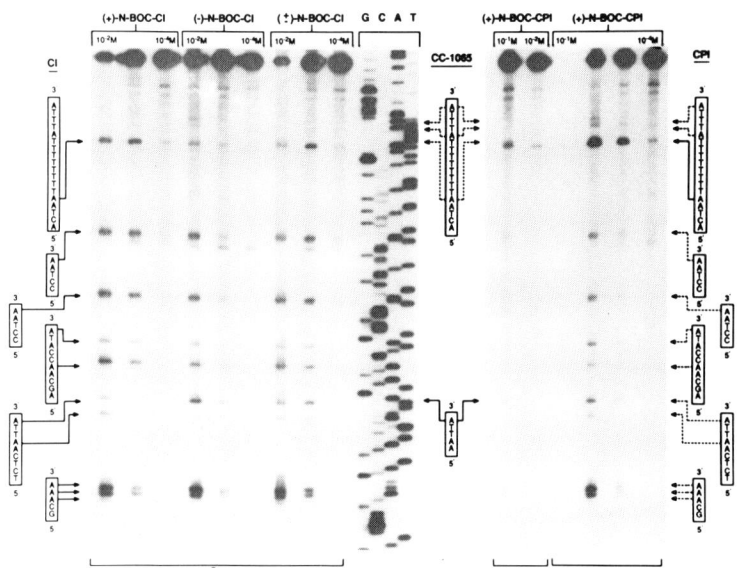

Figure 26. Thermally-induced strand cleavage of double-stranded DNA (SV40 DNA, 144 b.p.; nucleotide # 138-5238, clone w794) after 24 h incubation of agent:DNA at 4°C or 37°C followed by removal of unbound agent and 30 min incubation at 100°C; 8% denaturing poly(acrylamide) gel and autoradiography. Lanes 1-3, (+)-N-BOC-CI (4°C, 1 x 10^{-2} - 1 x 10^{-4} M); Lane 4-6, (-)-N-BOC-CI (4°C, 1 x 10^{-2} - 1 x 10^{-4} M); Lane 7-9, (±)-N-BOC-CI (4°C, 1 x 10^{-2} - 1 x 10^{-4} M); Lane 10-13, Sanger G, C, A, and T reactions; Lane 14-15, (+)-N-BOC-CPI (4°C, 2.5 x 10^{-1} - 2.5 x 10^{-2} M); Lane 16-18, (+)-N-BOC-CPI (37°C, 2.5 x 10^{-1} - 2.5 x 10^{-4} M).

Figure 25 Comparative In Vitro Cytotoxic Activity

Agent	Configuration	IC_{50}, L1210 (10^{-6} g/mL)	
(+)-N-BOC-CI	natural	4	$t_{1/2}$ = 35 sec
(−)-N-BOC-CI	unnatural	22	pH = 3
(+)-CI-CDPI$_1$	natural	0.010	
(−)-CI-CDPI$_1$	unnatural	0.020	
(+)-CI-CDPI$_2$	natural	0.005	
(−)-CI-CDPI$_2$	unnatural	0.015	
(+)-CC-1065	natural	0.00001	
(+)-N-BOC-CPI	natural	0.1	$t_{1/2}$ = 37 h, pH = 3

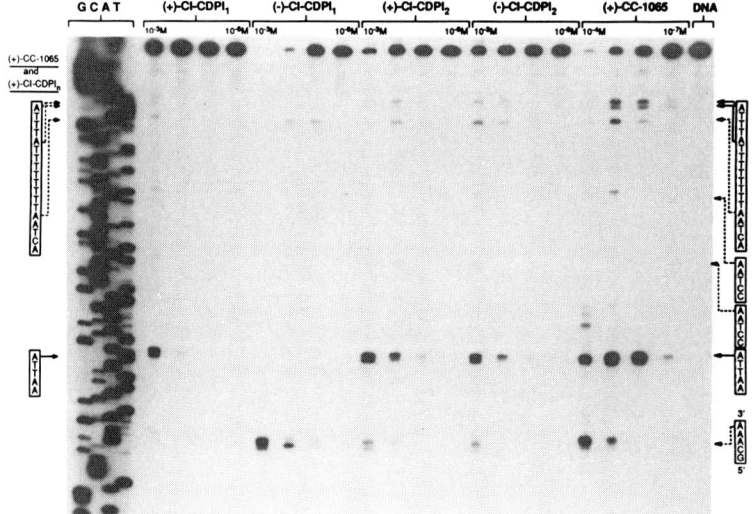

Figure 27. Thermally-induced strand cleavage of double-stranded DNA (SV40 fragment, 144 b.p.; nucleotide # 138-5238, clone w794) after 24 h incubation of agent:DNA at 4°C followed by removal of unbound agent and 30 min incubation at 100°C; 8% denaturing poly(acrylamide) gel and autoradiography. Lanes 1-4, Sanger G, C, A, and T reactions; Lanes 5-8, (+)-CI-CDPI$_1$ (5.6 x 10^{-4} - 5.6 x 10^{-7} M); Lanes 9-12, (−)-CI-CDPI$_1$ (5.6 x 10^{-4} - 5.6 x 10^{-7} M); Lanes 13-16, (+)-CI-CDPI$_2$ (5.6 x 10^{-4} - 5.6 x 10^{-7} M); Lanes 17-20, (−)-CI-CDPI$_2$ (5.6 x 10^{-4} - 5.6 x 10^{-7} M); Lanes 21-24, (+)-CC-1065 (1, 5.6 x 10^{-5} - 5.6 x 10^{-8} M); Lane 25, control DNA.

Functional Analogs of CC-1065　　123

electrophile alkylation selectivity and consequently we have interpreted the CC-1065 characteristic DNA alkylation selectivity of (+)-CI-CDPI$_2$ to represent a demonstration of the <u>dominant</u> effect of the agents noncovalent binding selectivity controlling its DNA alkylation selectivity. That is, <u>the preferential binding of the agents within the narrower, sterically more accessible A-T rich minor groove effectively restricts the number of accessible adenine N-3 alkylation sites for even the exceptionally reactive CI electrophile.</u>

7 ENHANCED FUNCTIONAL ANALOGS: (+)-CBI-CDPI$_n$

The establishment that the introduction of structural changes that increased the reactivity of the alkylation subunit of the agents (CI-based agents) was accompanied by a diminished intrinsic DNA covalent alkylation selectivity and DNA covalent alkylation intensity that proved directly related to the agents diminished cytotoxic potency suggested that the introduction of structural changes that would decrease the reactivity of the alkylation subunit (<u>increased</u> stability) would potentially enhance the agents productive DNA alkylation selectivity and intensity and enhance the resultant cytotoxic and antitumor potency of the agents. The first such agent examined to date is illustrated in Figure 28 where the 5-membered pyrrole ring of the alkylation subunit of CC-1065 has been replaced with a 6-membered benzene ring. The intuitive rationale suggesting the increased chemical stability of the CBI-based agents was based on an anticipated relative relief of strain introduced by the fused 6-membered <u>versus</u> 5-membered aromatic ring, was supported through computational studies, Figure 28, and ultimately confirmed upon experimental examination of the agents.

Figure 28

An Enhanced Functional Analog: (+)-CBI-CDPI$_2$

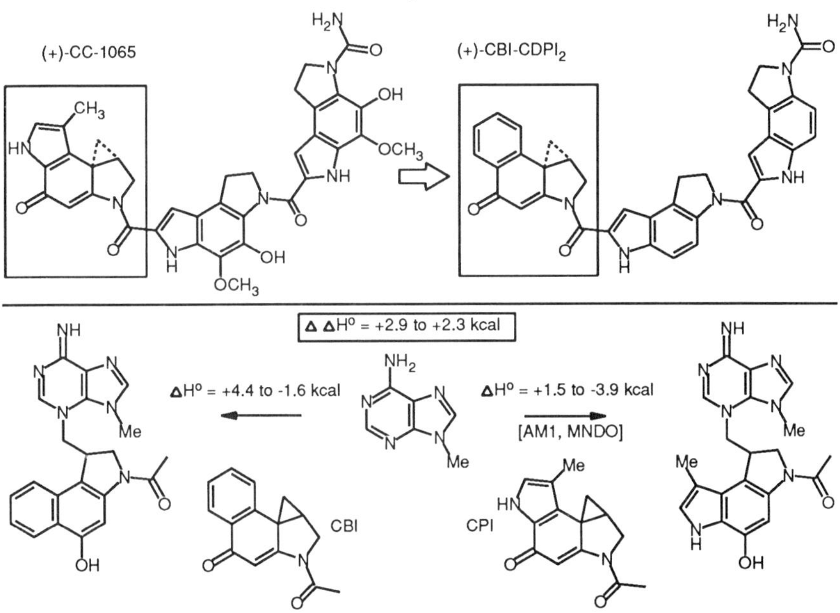

Figure 29

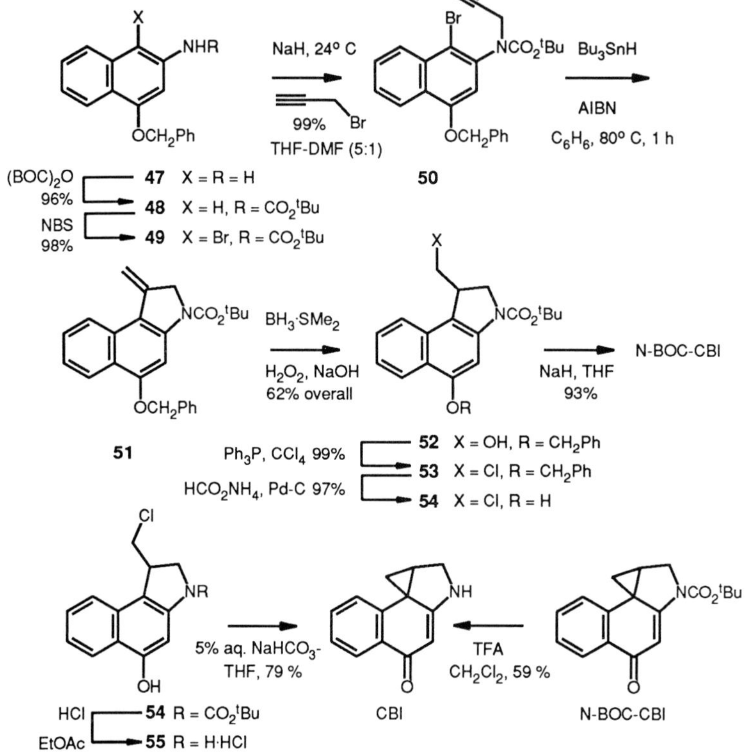

Figure 30

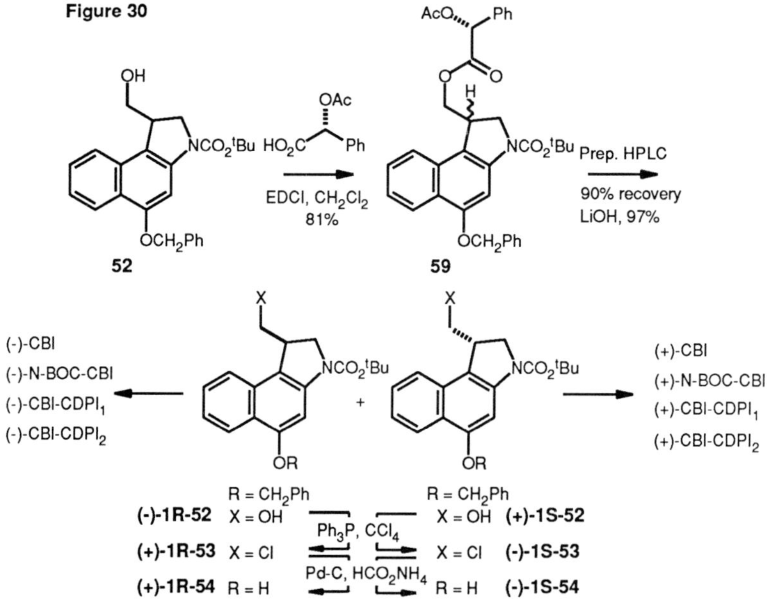

Figure 31

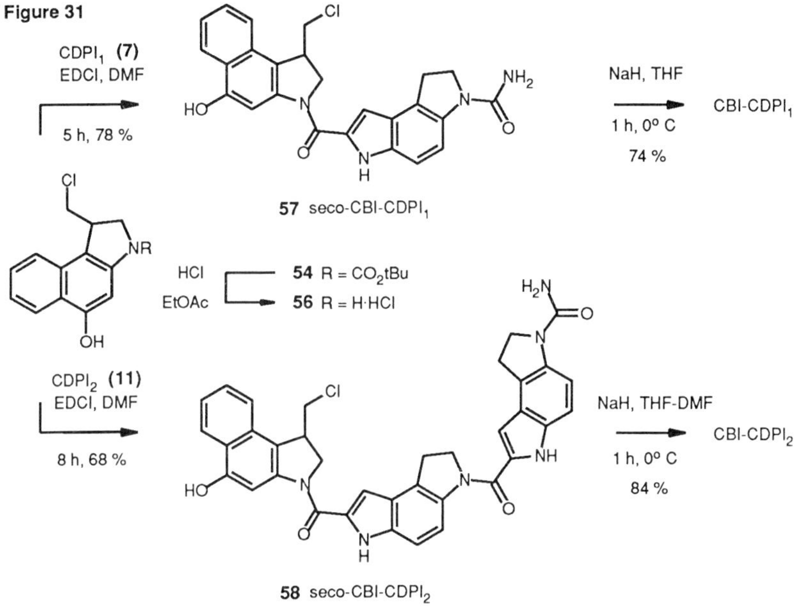
CDPI1 (7)
EDCI, DMF
5 h, 78 %
57 seco-CBI-CDPI1
NaH, THF
1 h, 0° C
74 %
CBI-CDPI1
HCl
EtOAc
54 R = CO2tBu
56 R = H·HCl
CDPI2 (11)
EDCI, DMF
8 h, 68 %
58 seco-CBI-CDPI2
NaH, THF-DMF
1 h, 0° C
84 %
CBI-CDPI2

The precursor to the CBI subunit was prepared as detailed in Figure 29[38] following closely the approach developed for the preparation of the CPI-based agents and relied on a key 5-exo-dig aryl radical-alkyne free radical cyclization. Surprisingly, the parent CBI agent could be prepared by acid-catalyzed (TFA) deprotection of N-BOC-CBI in acceptable conversions without preferential solvolysis of the activated cyclopropane and attests to the chemical stability of the agents and the modest electrophilic character they possess.

The resolution of a CBI precursor through chromatographic separation of diastereomers **59**, Figure 30,[39] and the incorporation of the resolved enantiomers of **52** into the preparation of the optically active CBI-based agents followed the protocols presented in Figures 29 and 31. The unambiguous assignment of the absolute configuration was achieved through single crystal X-ray analysis of (-)-1S-**53**.[39]

Figure 32 Comparative In Vitro Cytotoxic Activity

Agent	IC_{50} (10^{-11} g/mL, pM)		
(+)-N-BOC-CBI	2300	77000	$t_{1/2}$ = 129 h, pH = 3
(-)-N-BOC-CBI	28000	940000	
(+)-CBI-CDPI$_1$	0.2	5	
(-)-CBI-CDPI$_1$	> 160	> 380	
(+)-CBI-CDPI$_2$	0.3	5	
(-)-CBI-CDPI$_2$	3	50	
(+)-CC-1065	1.2	20	$t_{1/2}$ = 37 h, pH = 3
(+)-N-BOC-CPI	10000	330000	

Figure 33

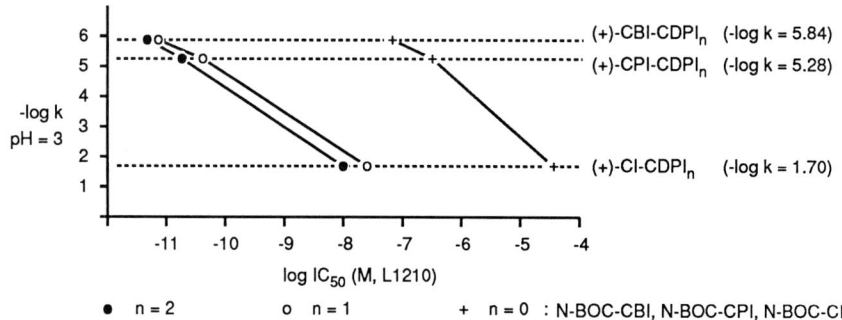

- n = 2   o n = 1   + n = 0 : N-BOC-CBI, N-BOC-CPI, N-BOC-Cl

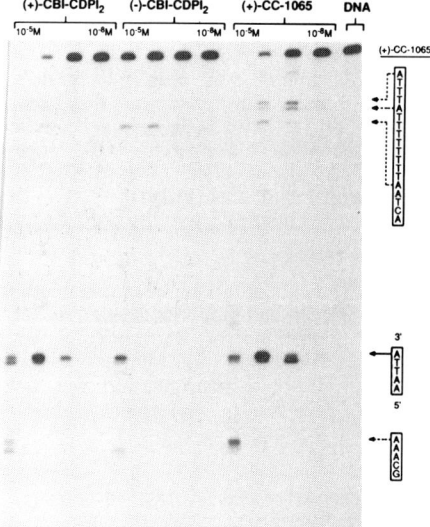

Figure 34. Thermally-induced strand cleavage of a double-stranded SV40 DNA fragment (144 b.p.; nucleotide # 138-5238, clone w794) after 24 h incubation of agent:DNA at 4°C followed by removal of unbound agent and 30 min incubation at 100°C; 8% denaturing poly(acrylamide) gel electrophoresis, and autoradiography. Lanes 1-4, (+)-CBI-CDPI$_2$ (1 x 10^{-5} - 1 x 10^{-8} M); Lanes 5-8, (-)-CBI-CDPI$_2$ (1 x 10^{-5} - 1 x 10^{-8} M); Lanes 9-12, (+)-CC-1065 (1 x 10^{-5} - 1 x 10^{-8} M); Lane 13, control DNA.

Consistent with expectations, N-BOC-CBI ($t_{1/2}$ = 129 h, pH = 3) proved substantially more stable (ca. 4x) than N-BOC-CPI ($t_{1/2}$ = 37 h, pH = 3) to acid-catalyzed solvolysis and accompanying this increased stability was a substantial increase in the cytotoxic potency of the agents (ca. 4x) relative to that of the CPI-based agents, Figure 32. The cytotoxic potency of CBI-CDPI$_1$ and CBI-CDPI$_2$ proved remarkably high (5 pM, 2-3 pg/mL), nearly indistinguishable from one another, and comparable in potency to the most potent cytotoxic agents identified to date including the esperamycin/calicheamicin agents. Like observations made in the study of the CPI-based agents, the 8bR,9aS enantiomers that possess the absolute configuration of natural (+)-CC-1065 proved consistently more potent than the unnatural enantiomers and the relative cytotoxic potency of the agents follow the relative intensity and selectivity trends established in DNA alkylation studies, Figure 34. Although our examination of the agents is not yet complete, the CBI-based agents appear to be slightly more sequence selective (discriminate) than the CPI-based agents and exhibit a comparable or slightly more intense DNA covalent alkylation. The enhanced cytotoxic potency of (+)-CBI-CDPI$_1$ and (+)-CBI-CDPI$_2$ relative to that of (+)-CPI-CDPI$_1$/(+)-CPI-CDPI$_2$/(+)-CC-1065 is not in agreement with the proposed relationship that the agents productive DNA binding properties and resulting expression of cytotoxic potency are related directly to the reactivity of the electrophile as extrapolated from acid-catalyzed solvolysis rates and directly related to the rate of acid-catalyzed covalent alkylation of DNA.[4,6] In fact, the comparison of the cytotoxic potency of (+)-CI-CDPI$_n$, (+)-CPI-CDPI$_n$, and (+)-CBI-CDPI$_n$ and the corresponding rates of acid-catalyzed solvolysis of N-BOC-CI, N-BOC-CPI, and N-BOC-CBI (pH = 3) suggests that the inverse relationship between electrophile reactivity and cytotoxic potency may constitute a more relevant relationship, Figure 33. This presumably results from more selective and productive (agent availability) covalent modification of DNA although the precise origin of this relationship is under investigation.

8 ACKNOWLEDGMENTS

It is my pleasure to acknowledge the extraordinary efforts of a spirited group of colleagues that have been responsible for the conduct of the studies described: R.S. Coleman, R.J. Wysocki, Jr., S.M. Sakya, Dr. B.J. Invergo, H. Zarrinmayeh, Dr. T. Ishizaki, Dr. S.A. Munk, and to provide a special acknowledgment to our collaborators: Professor P.A. Kitos (University of Kansas, Department of Biochemistry), S.C. Thompson, and O. Suntornwat for securing the cytotoxic testing results detailed herein. This work was assisted financially by the National Institutes of Health (CA 41986).

9 REFERENCES

1. L.J. Hanka, A. Dietz, S.A. Gerpheide, S.L. Kuentzel, and D.G. Martin, J. Antibiot., 1978, 31, 1211.

2. L.H. Hurley and D.R. Needham-VanDevanter, Acc. Chem. Res., 1986, 19, 230.

3. V.L. Reynolds, J.P. McGovren, and L.H. Hurley, J. Antibiot., 1986, 39, 319.

4. M.A. Warpehoski and L.H. Hurley, Chem. Res. Toxicol., 1988, 1, 315.

5. W. Wierenga, B.K. Bhuyan, R.C. Kelly, W.C. Krueger, L.H. Li, J.P. McGovren, D.H. Swenson, and M.A. Warpehoski, Adv. Enz. Regul., 1986, 25, 141.

6. M.A. Warpehoski, I. Gebhard, R.C. Kelly, W.C. Krueger, L.H. Li, J.P. McGovren, M.D. Prairie, N. Wicnienski, and W. Wierenga, J. Med. Chem., 1988, 31, 590.

7. L.H. Hurley, V.L. Reynolds, D.H. Swenson, G.L. Petzold, and T.A. Scahill, Science, 1984, 226, 843.

8. L.H. Li, T.L. Wallace, T.F. DeKoning, M.A. Warpehoski, R.C. Kelly, M.D. Prairie, and W.C. Krueger, Invest. New Drugs, 1987, 5, 329.

9. V.L. Reynolds, I.J. Molineaux, D.J. Kaplan, D.H. Swenson, and L.H. Hurley, Biochemistry, 1985, 24, 6228.

10. L.H. Hurley, C.-S. Lee, J.P. McGovren, M.A. Warpehoski, M.A. Mitchell, R. C. Kelly, and P.A. Aristoff, Biochemistry, 1988, 27, 3886.

11. C.G. Chidester, W.C. Krueger, S.A. Mizsak, D.J. Duchamp, and D.G. Martin, J. Am. Chem. Soc., 1981, 103, 7629.

12. R.S. Coleman and D.L. Boger, "Studies in Natural Products Chemistry", Atta-ur-Rahman, ed., Elsevier, Amsterdam, 1989, Vol. 3, p. 301.

13. V.H. Rawal, R.J. Jones, and M.P. Cava, Heterocycl., 1987, 25, 701.

14. R.J. Sundberg and E.W. Baxter, Tetrahedron Lett., 1986, 27, 2687.

15. R.J. Sundberg, E.W. Baxter, W.J. Pitts, R. Ahmed-Schofield, and T. Nishiguchi, J. Org. Chem., 1988, 53, 5097.

16. K.J. Drost, R.J. Jones, and M.P. Cava, J. Org. Chem., 1989, 54, 5985.

17. T.A. Bryson and G.A. Roth, Tetrahedron Lett., 1988, 29, 2167.

18. T.A. Bryson, G.A. Roth, and L. Jing-hau, Tetrahedron Lett., 1986, 27, 3685.

19. D.L. Boger and R.S. Coleman, J. Am. Chem. Soc., 1988, 110, 4796.

20. D.L. Boger and R.S. Coleman, J. Am. Chem. Soc., 1988, 110, 1321.

21. D.L. Boger and R.S. Coleman, J. Am. Chem. Soc., 1987, 109, 2717.

22. D.L. Boger and R.S. Coleman, Tetrahedron Lett., 1987, 28, 1027.

23. D.L. Boger and R.S. Coleman, J. Org. Chem., 1986, 51, 3250.

24. D.L. Boger, R.S. Coleman, B.J. Invergo, H. Zarrinmayeh, P.A. Kitos, S.C. Thompson, T. Leong, and L.W. McLaughlin, Chem.-Biol. Interactions, 1990, 73, 29.

25. D.L. Boger, R.S. Coleman, and B.J. Invergo, J. Org. Chem., 1987, 52, 1521.

26. D.L. Boger and R.S. Coleman, J. Org. Chem., 1984, 49, 2240.

27. M. Coll, J. Aymami, G. A. van der Marel, J.H. van Boom, A. Rich, and A.H.-J. Wang, Biochemistry, 1989, 28, 310.

28. M. Coll, C.A. Frederick, A.H.-J. Wang, and A. Rich, Proc. Natl. Acad. Sci. (USA), 1987, 84, 8385.

29. R. Wing, H. Drew, T. Takano, C. Broka, S. Tanaka, K. Itakura, and R.E. Dickerson, Nature (London), 1980, 287, 755.

30. H.C.M. Nelson, J.T. Finch, B.F. Luisi, and A. Klug, Nature (London), 1987, 330, 221.

31. M.A. Warpehoski and V.S. Bradford, Tetrahedron Lett., 1988, 29, 131.

32. D.L. Boger and R.S. Coleman, J. Org. Chem., 1988, 53, 695.

33. D.L. Boger, R.S. Coleman, B.J. Invergo, S.M. Sakya, T. Ishizaki, S.A. Munk, H. Zarrinmayeh, P.A. Kitos, and S.C. Thompson, J. Am. Chem. Soc., 1990, 112, in press.

34. D.L. Boger and S.A. Munk, unpublished studies.

35. C. Ambrose, A. Rajadhyaksha, H. Lowman, and M. Bina, J. Mol. Biol., 1989, 210, 255.

36. D.L. Boger and R.J. Wysocki, Jr., J. Org. Chem., 1989, 54, 1238.

37. D.L. Boger, R.J. Wysocki, Jr., and T. Ishizaki, J. Am. Chem. Soc., 1990, 112, in press.

38. D.L. Boger, T. Ishizaki, R.J. Wysocki, Jr., S.A. Munk, P.A. Kitos, and O. Suntornwat, J. Am. Chem. Soc., 1989, 111, 6461.

39. D.L. Boger and T. Ishizaki, Tetrahedron Lett., 1990, 31, 793.

40. D.L. Boger and H. Zarrinmayeh, J. Org. Chem., 1990, 55, 1379.

Stereochemical Effects During Synthesis of New Heterocycles via Intramolecular Dipolar Cycloadditions

A. Hassner

DEPARTMENT OF CHEMISTRY, BAR-ILAN UNIVERSITY, RAMAT-GAN 52100, ISRAEL

1,3-Dipolar cycloadditions have been of wide interest both from the synthetic and the theoretical point of view. These reactions were extensively developed by Huisgen and coworkers (Angew.Chem.Int.Ed. **1963**, *2*, 565) and by many other research groups. In as much as they involve ring formation via 4+2 cycloadditions, they constitute the hetero analog of the Diels-Alder reaction. A number of dipoles shown below have been studied and can react with different multiple bond dienophiles to produce 5-membered ring heterocycles.

1,3-Dipolar Cycloadditions

Huisgen

$$\overset{Y+}{X}\diagdown_{Z^-} + \diagup R \qquad X\overset{Y}{=}Z\diagdown_R$$

1,3-dipole   dipolarophile

examples of dipoles:

$$R-C\equiv\overset{+}{N}-\overset{-}{O} \quad R-C\equiv\overset{+}{N}-\overset{-}{C} \quad R-\overset{-}{N}-\overset{+}{N}\equiv N$$

$$R-C=\underset{+}{\overset{R}{N}}-\overset{-}{O} \quad R-C=\underset{+}{\overset{R}{N}}-\overset{-}{C}$$

Such heterocyclic rings can serve as a useful entry into functionalized systems. For instance, the nitrile oxide dipole, shown below, obtainable from aldoximes (with NaOCl or similar reagents) or from primary nitro compounds (by reaction with aromatic isocyanates), has been studied extensively by the groups of Quilico, Huisgen and others and was shown to react with terminal alkenes to form primarily 3,5-disubstituted isoxazolines. More recently, Kozikowski, Jager, Curran and others have demonstrated the usefulness of such isoxazolines in the synthesis of ß-hydroxy ketones or of 1,3-amino alcohols.

Stereochemical Effects During Synthesis of New Heterocycles

We have been interested to achieve and predict regio- and stereoselectivity during intramolecular cycloadditions of nitrile oxides with olefins (INOC), with a view towards simultaneous double annulation of two rings. One ring would be heterocyclic(or carbocyclic), the second an isoxazoline which can be opened to produce stereoselectively positioned functional groups. Such intramolecular reactions, had been less well explored than their intermolecular counterparts and the stereochemical features guiding the cycloadditions are still not too well understood in spite of the pioneering work of Kozikowski and coworkers (Acc. Chem. Res. **1984**, *17*, 410).

An application of the INOC reaction to the synthesis of Ptilocaulin **1**, an antileukemic agent isolated by Rheinhardt et al. from a marine sponge found in the Carribean Sea is illustrated below. Ptilocaulin possesses four stereocenters of which three are contiguously cis. A retrosynthetic analysis, via a keto alcohol, shown below provided a possible approach by means of cycloaddition of the unsaturated nitrile oxide **a**, in the hope that during the cycloaddition all three centers will be introduced cis. This indeed later proved to be case.

As an entry into the nitrile oxide we chose the aldoxime route. Recently, we showed that Z-aldoximes can be converted into their dianions, which can undergo aldol condensation with ketones in good yield (Chem.Ber. **1988**, *121*, 1823). Thus, we were able to condense the Z-hexanal oxime dianion with acetonylcyclopentene in high yield. This provided **2** (a mixture of isomers) with a handle for the introduction of the fourth stereocenter at a later step. Reaction of the aldoxime **2** with NaOCl led to a nitrile oxide which underwent spontaneous cyclization to the isoxazoline **3**, as a mixture of isomers. That the cycloaddition had taken place stereospecifically was demonstrated when elimination of water (by means of SOCl$_2$) yielded **4** as a single stereoisomer. N-O cleavage with Ra-Ni produced the ß-hydroxy ketone **5**. Introduction of the ß-methyl stereochemistry in **6** proved difficult but was finally achieved by means of Li-EtNH$_2$. For the formation of ptilocaulin **1** it proved necessary to first convert **6** into the unsaturated ketone **7** which underwent a Michael addition-condensation with guanidine to yield **1**.

While HOMO-LUMO interactions normally direct intermolecular nitrile oxide additions to 1-alkenes to produce regioselectively 5-substituted isoxazolines, intramolecular additions appear to behave differently. The most important question to be answered was that of stereochemistry during ring formation.

If one can take advantage of regio and stereoselectivity during intramolecular dipolar cycloadditions, one shoud be able to apply such reactions to the synthesis of carbopenams or to lupine or necine alkaloids. The importance of being able to assess both regiochemical and stereochemical preferences during such dipolar cycloadditions is illustrated in the retrosynthetic analysis leading for isoretronecanol.

We first investigated the readily available 2-vinylazetidinones as model systems. In order to introduce a nitrile oxide side chain we monoalkylated the ß-lactam with dibromoalkanes of variable chain length in the presence of solid KOH. The resulting bromo derivatives were converted to nitro compounds which reacted with PhNCO - Et$_3$N to produce nitrile oxides that cyclized spontaneously.

When a 6-membered ring was formed during the double hetero annulation, a single isomer was produced in over 90% yield. NOE indicated cis stereochemistry. On the other hand when a 5- or 7-membered ring fused to the isoxazoline was formed, then a mixture of cis and trans isomers resulted.

This phenomenon can be explained by a preference of a chair over a boat transition state during the cyclization to the 6-membered ring. In the more flexible 5- or 7-membered ring transition states the difference in energy leading to cis vs trans isomers should be less pronounced. Indeed, MM2 transition state energy calculations (by Prof. A. Padwa) indicated a difference of 1.5 kcal in favor of the cis isomer during formation of the 6-membered ring (3 kcal for ground state energy difference), while there was only a difference of ca. 0.5 kcal during ring closure to the 7-membered ring compounds.

Chair Transition State   cis
Et=43.13 kcal

Boat Transition State   trans
Et=44.67 kcal

In order to test the generality of this stereochemical preference we prepared the analogous azides (in quantitative yield, by reaction of the bromoalkylazetidinones with a polymeric azide reagent, Angew.Chem.Int.Ed. **1986**, 25, 478, J.Org.Chem. **1990**, 55, 2304). Heating of the azido olefins again produced a single isomer (cis) during ring closure to a 6-membered ring while in all other cases a mixture of cis and trans isomers resulted.

$$\text{P}-\overset{+}{N}R_3 \ \overset{-}{Cl} + Na^+N_3^- \longrightarrow \text{P}-\overset{+}{N}R_3 \ N_3^- \ \underline{1}$$
polymeric reagent

$$R-X + \underline{1} \xrightarrow[\text{any solvent}]{20°} R-N_3 \ + \text{P}-\overset{+}{N}R_3 \ X^-$$
100%   filter

X: I, Br, Cl, OSO_2R

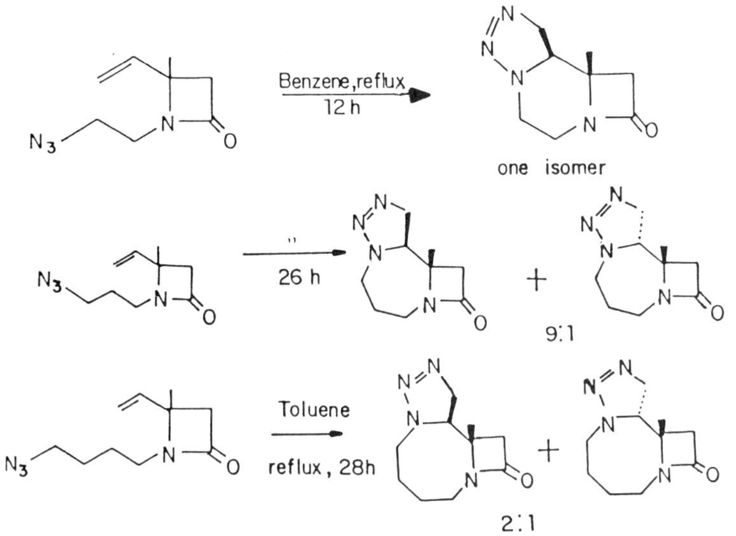

These novel ß-lactam triazolines undergo further stereospecific transformations on silica gel.

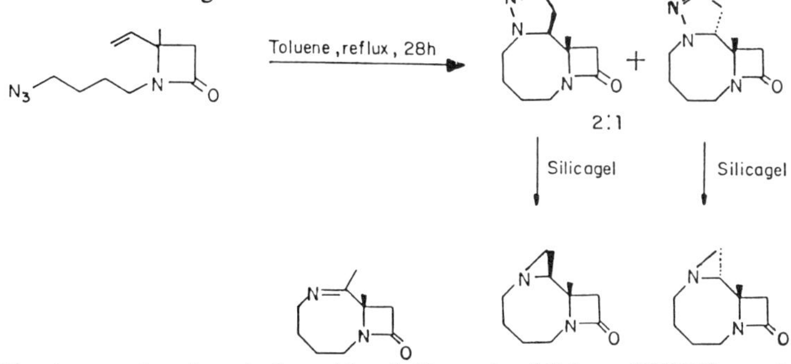

The intramolecular nitrile oxide olefin cycloadditions (INOC) can be extended to involve the double bond of a pyrrole or of an indole, as shown below. These cycloadditions produced 6- or 7-membered ring fused heterocycles but were not successful in fusing a 5-membered ring to the pyrrole or indole.

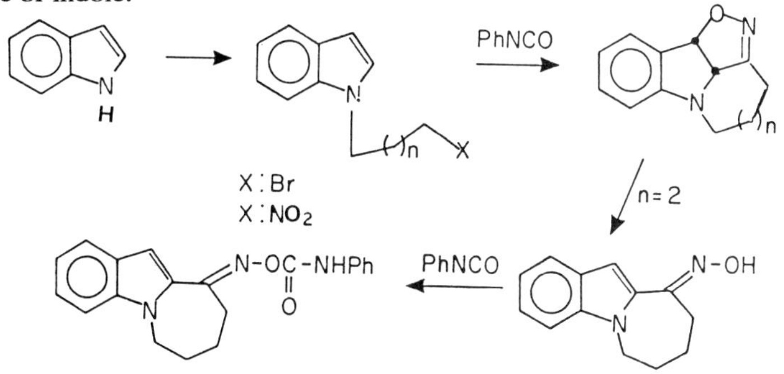

In a similar manner, sulfonamides of allylamines can be converted into nitrile oxide or azides. These compounds serve as starting points for formation of functionalized N-heterocycles containing 5,6,7- or 8-membered rings, via thermal or free radical cyclizations.

Vinylazetidines were not amenable to the above treatment, hence we developed a general method to introduce a nitrile oxide side chain via α-bromoaldoximes. Unfortunately, the latter compounds unlike their keto analogues are difffficult to prepare and unstable.

Stereochemical Effects During Synthesis of New Heterocycles

[Scheme showing vinylazetidine transformations: HN-containing azetidine with vinyl group → HO-N=CH-N-containing bicyclic structure with R group]

R-CH₂-C(H)=N-OH →(Br₂) R-CH(Br)-CH=NOH

R-CH₂-C(H)=N-OH →(poor) R-CH(Br)-CH=O

We successfully circumvented this difficulty by protecting the aldoxime as its silyl ether, which now allowed alpha bromination with NBS to take place. The α-bromo-O-silylaldoximes have another advantage, namely they can be treated with fluoride ions and converted into a very reactive vinyl nitroso species, which is immediately trapped by nucleophiles present.

$$R\text{-}CH_2\text{-}CH=N\text{-}OH \xrightarrow{Me_3SiCl} R\text{-}CH_2\text{-}C(H)=N\text{-}O\,SiMe_3 \xrightarrow{NBS} R\text{-}CH(Br)\text{-}C(H)=N\text{-}O\,SiMe_3$$

$$\downarrow X^-$$

$$R\text{-}CH(X)\text{-}CH=N\text{-}O\,SiMe_3$$

$$\downarrow F^-$$

$$R\text{-}CH=CH\text{-}N=O$$

$$\downarrow F^-, X^-$$

$$R\text{-}CH(X)\text{-}CH=N\text{-}OH$$

In this manner it is possible to attach a nitrile oxide side chain onto a vinylazetidine and study the stereochemistry of the dipolar cycloaddition. Only two of the possible four isomers were isolated in this ring closure. NMR indicated the trans,cis and cis,cis structures (in a ratio of 2:1, R = Ph). Transition state models indicate more unfavorable interactions (allylic) leading to the minor cis,cis (than to the trans,cis) product. Furthermore, models indicate that strong repulsive interactions should result if R is alpha rather than beta. Indeed, no alpha phenyl product was found and MM2 calculations again confirm these results indicating a 1.6 kcal difference between the trans,cis and the cis,cis isomer, with both of these being more stable than any isomer in which R is alpha.

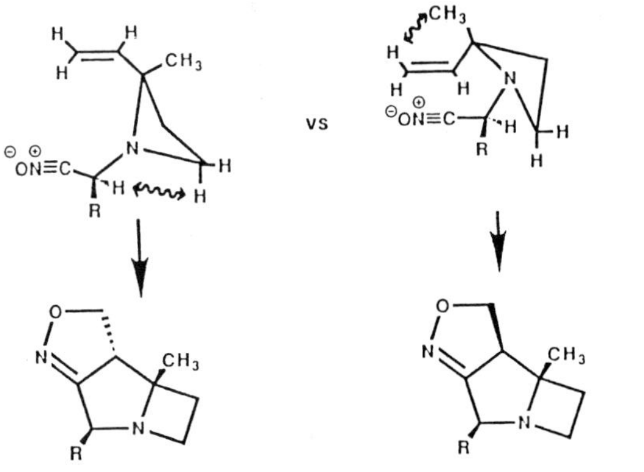

trans,cis (67%)
Et=44.17 kcal

cis,cis (33%)
Et=45.59 kcal

It is possible to extend this methodology via the bromo-O-silylaldoxime route, for the general conversion of an unsaturated chain, containing a nucleophilic center such as an amine, into an oxime olefin which serves as a substrate for INOC.

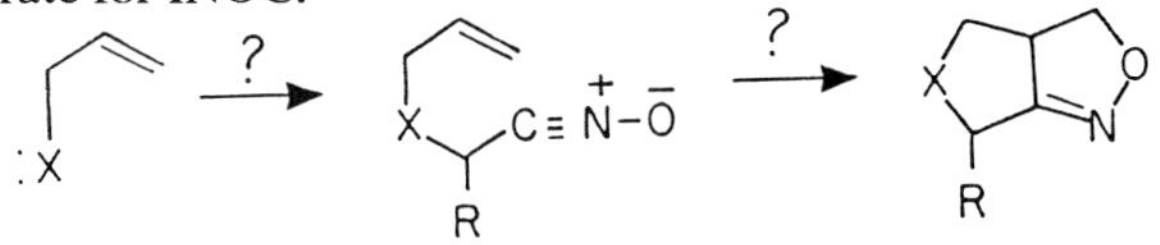

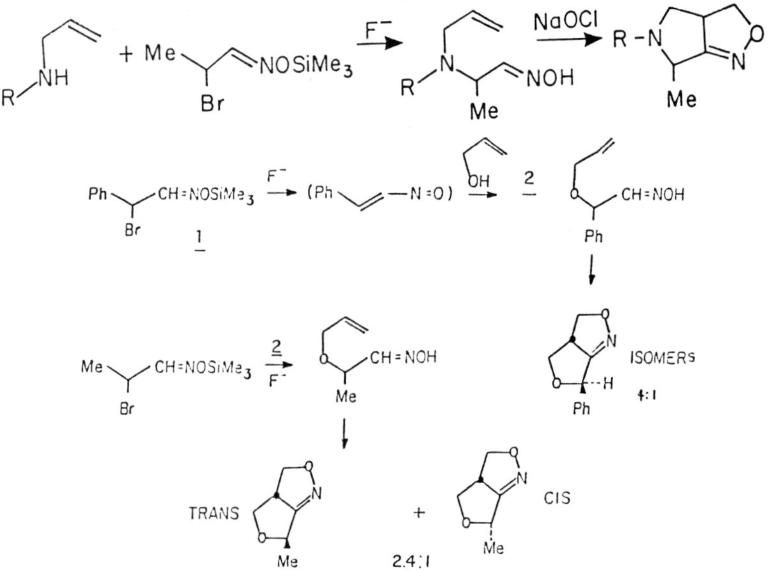

Unsaturated alcohols can likewise be converted by this method to unsaturated oximes, which undergo the cycloaddition (see above). The stereochemistry of the resulting fused tertrahydrofurans depends on the nature of the substituent on the newly formed furan ring, but usually a trans :cis mixture results with the trans predominating. MM2 calculations indicate only a small difference between the trans and cis isomers in the cyclic ether series.

Functionalized tetrahydropyrans and other cyclic ethers can be formed in a similar manner.

Application of this methodolgy to the formation of fused tetrahydrothiophenes was not successful because of the sensitivity of the sulfide to the chlorinating agents employed.

Hence, we devised a different approach via nitro olefins. It was in fact possible (see above) to achieve formation of the required unsaturated olefins **e** in a one pot reaction from aldehydes, nitromethane and an unsaturated thiol.

This led us to a different approach to unsaturated nitrile oxide ethers **f** (X = O). Instead of starting with aldehydes of type RCH_2CHO and proceeding to **f** via vinyl nitroso intermediates, we converted the lower homolog RCHO into the nitrile oxides **f** via vinyl nitro compounds.

Furthermore, it was possible to carry out a dipolar cycloaddition starting from the nitro olefins either via an intramolecular nitrile oxide olefin cycloaddition (INOC) or via an intramolecular silyl nitronate olefin cycloaddition (ISOC). The latter route had the advantage of essentially complete stereospecifity.

Silyl nitronates

This was especially noticeable during the formation of isoxazolothiophenes, where nitrile oxides led to a 1:1 mixture of stereoisomers while the silyl nitronate route provided only the trans isomer in good yield.

tetrahydrothiophenes

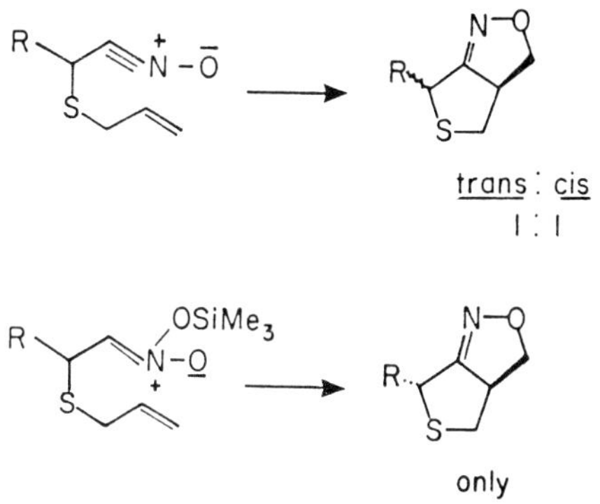

trans : cis
1 : 1

only

The table provides a comparison of the dramatic differences in stereoselectivity that can be achieved during intramolecular dipolar cycloadditions via nitrile oxide olefins (INOC) vs those via silyl nitronate olefins (ISOC) for functionalized tetrahydrofuran, tetrahydro-thiophene and even cyclopentane systems.

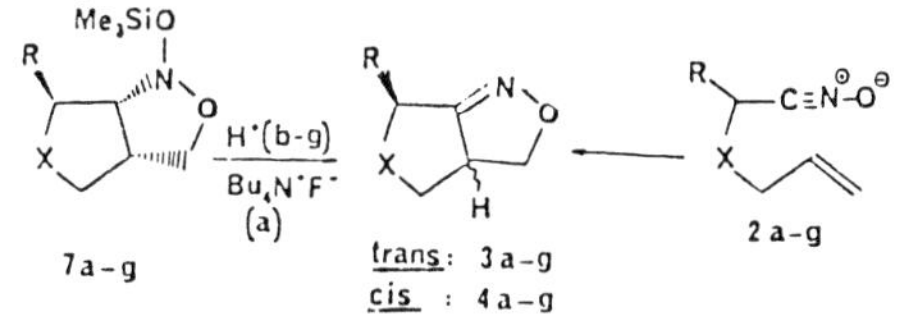

	R	X	INOC reaction		ISOC reaction	
			yield %	ratio 3:4	yield %	ratio 3:4
a	Me	O	87	2.4:1	74	>99:1
b	Me	S	84	1:1	89	>99:1
c	iPr	S	90	1:1	84	>99:1
d	Ph	S	80	3:2	85	>99:1
e	4-MeO-C$_6$H$_4$	S	63	3:2	87	>99:1
f	Me	C(COOMe)$_2$	81	1:7	84	>99:1
g	4-MeO-C$_6$H$_4$	C(COOMe)$_2$	90	1:5	91	>99:1

Another cycloaddition mode that proved highly stereoselective is the intramolecular oxime olefin cycloaddition (IOOC). During our studies with the unsaturated oxime **g** (below), which upon treatment with NaOCl gave an isoxazolo pyrrolidine via an INOC reaction, we attempted to heat **g** and obtained in good yield a fused isoxazolidine as a single isomer (cis,trans).

We believe that this reaction proceeds via an H-nitrone, a species that has so far eluded isolation and that is formed in equilibrium concentrations from the oxime by a proton shift from O to N. The new route provides a superior stereoselective synthesis of amino alcohol functionalized pyrrolidines.

Indeed Grigg and coworkers (Chem. Soc.Rev. 1987, 16, 89) have recently shown that oximes can undergo Michael addition to acceptor olefins followed by cycloaddition of the intermediate nitrone to the olefin.

We have used these oxime olefin cycloadditions in the stereospecific formation of pyrrolidines containing 3 to 4 stereocenters.

Stereochemical Effects During Synthesis of New Heterocycles

[Scheme showing nitrone formation from oxime and thermal cyclization to give two isomeric bicyclic isoxazolidines: "1 isomer, 4 stereocenters" (with Et, Ph, Me substituents) and "1 isomer, 3 stereocenters" (with Et).]

Furthermore, these cyclizations can be used in the stereoselective synthesis of pyrrolizidines related to necine alkaloids, as well as indolizidines or quinolizidines.

[Scheme: proline → 2-vinylpyrrolidine → N-alkylated vinyl pyrrolidine with C=NOH → bicyclic isoxazolidine-fused pyrrolizidine → pyrrolizidine diol/amine (OH, NH₂).]

By means of molecular mechanics calculations it is possible to predict the more stable isomer formed (for instance anti,cis for the indolizidines and syn,cis for the quinolizidines shown) as well as to correlate or predict NMR coupling constants.

Predict J values

[Scheme for indolizidine nitrone cyclization:]

anti J_{AB} 8.5 Hz
calcd anti 7.8
syn 5.3

[Scheme for quinolizidine nitrone cyclization:]

syn J_{AB} 2.8 Hz
calcd syn 3.2
anti 10.4

I am indebted to my coworkers Dr. K. Murthy, Dr. R. Maurya and Dr. W. Dehaen for their skill and devotion in this research, to Prof. A. Padwa for calculations and fruitful collaboration and to the US-Israel Binational Science Foundation for financial support.

New Reactions of Partially Hydrogenated Indeno[1,2-*b*]-Pyridines

G. Duburs, D. Muceniece, and V. Lusis

INSTITUTE OF ORGANIC SYNTHESIS, LATVIAN ACADEMY OF SCIENCES, RIGA, LATVIA

INTRODUCTION

1,4 -Dihydropyridine derivatives have attracted considerable attention largely on account of their ability to influence biochemical processes and because of a variety of practical applications.

The chemistry of 1,4-dihydropyridines has mainly developed in two directions including the synthesis of new derivatives of dihydropyridines aimed at obtaining biologically active compounds, and the use of synthetic 1,4-dihydropyridines as models for NADH coenzyme and investigations of hydride transfer processes with dihydropyridines.

In comparison with 1,4-dihydropyridines, the synthesis and reactions of indeno[1,2-b]-pyridines containing a hydrogenated pyridine ring have hardly received attention. The main results of our investigations of partially hydrogenated indeno[1,2-b]pyridines are briefly outlined in this review.

5-Oxo-1H-4,5-dihydroindeno[1,2-b]pyridines

The preparation of 5-oxo-1$\underline{H}$-4,5-dihydroindeno[1,2-b]pyridines **1** does not pose any particular problem and a comparatively large number of compounds are known bearing different substituents in positions 3 and 4 of the pyridine ring. The sequence of atoms N-C_{9b}=C_{4a}-C=O in the molecule of 5-oxo-4,5-di-hydroindenopyridines has been confirmed by means of IR spectroscopy as well as by specific reactions. 5-Oxo-1$\underline{H}$-4,5-dihydroindeno[1,2-b]pyridine derivatives **1** in the pre-sence of alkaline agents form anions with several centers of different hardness. During interaction with a hard alkylation agent (DMS or dimethyltosylate) the hard reaction centers take part in the reaction resulting exclusively in the indenopyridines **2** and **3**, the C-alkyl derivative **3** being the main product (*Scheme 1*). O-alkylation was not observed in this case.

The effect of solvents on the ratio of N- and C-alkylation products was investigated during the alkylation of 2-methyl-3-ethoxycarbonyl-4-phenyl-5-oxo-1H-4,5-dihydroindeno[1,2-b]-pyridine (**1**, R=CO$_2$Et) with methyl iodide. In aprotic solvents the ratio of C- and N-alkylation products (**3**:**2**) varied within the range 1.1-3.0. A protondonating me-dium significantly increased the amount of C-alkyl derivative (**3**:**2**=10), allowing one to obtain 4a,5-dihydroindenopyridine derivative **3** with an angular alkyl substituent at C-4a.

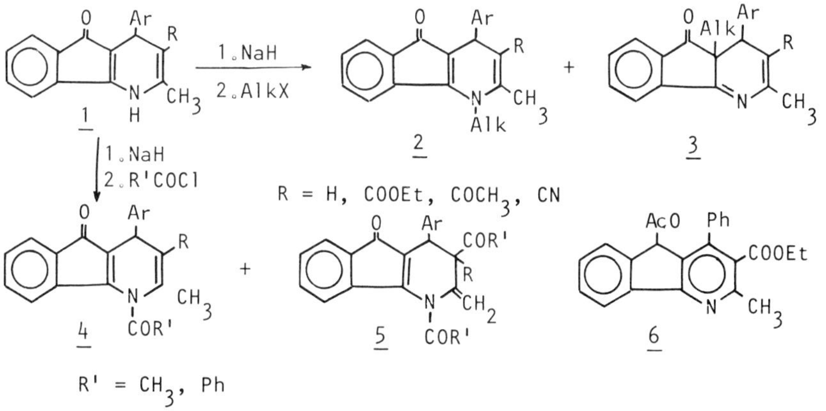

Scheme 1

The influence of protic solvents can be explained by the formation of a hydrogen bond between the dihydroindenopyridine anion and the solvent: the hard proton forms the strongest bond with the hardest center of the ambident anion and in this way sterically covers this center.

Anions of 5-oxo-4,5-dihydroindenopyridine derivatives in a non-polar aprotic medium react with acid chlorides to form mainly N-acyl-derivatives of dihydroindenopyridine **4**. In polar medium the N,C-diacyl derivatives **5** is formed together with the N-monoacyl derivatives **4** (*Scheme 1*). However it still remains unclear whether the N,C diacyl-ation products are formed via deprotonation of the 2-methylsubstituent of N-acyl derivative **4** or as a result of direct acylation of the enamino moiety in the molecule with [R-C=O]$^+$, according to the S_{n1} mechanism.

Formally, the hardest center of the N-C=C-C=O moiety is the oxygen atom, though the O-acylation of indenopyridines takes place only under phase-transfer conditions. The predominance of N-substituted derivatives in the acylation reaction and the absence of O-alkyl derivatives in the course of alkylation suggest that the negative charge of the indenopyridine anion is not uniformly distributed, a considerable amount of it is being located in the pyridine ring. That is why the acylation of the 5-CO group becomes possible only when the negative center, the nitrogen atom, is screened by the bulky crown-linked K^+ ions. Acetylation of undissociated NH-4,5-dihydroindenopyridine proceeds via acetylation of enaminoketone following the migrating of 4-H and affords the acetyl derivative **6**.

Another reaction involving the enaminoketonic moiety of 5-oxo-indenopyridine may be the exchange of the 5-CO group to thioxo-group. The derivatives of dihydroindenopyridines **1** and **2** with Lawesson's reagent undergo transformation into appropriate sulphur derivatives **7** under mild conditions[2], which is characteristic of unsaturated aminoketones (*Scheme 2*).

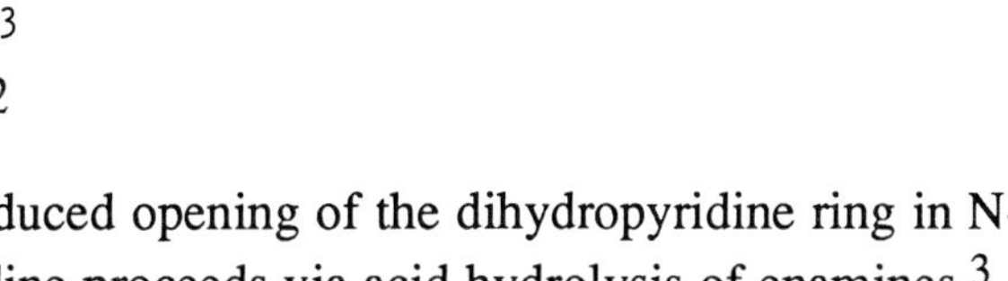

R = H, COOEt, COCH$_3$, CN
R' = H, CH$_3$

Scheme 2

Acid induced opening of the dihydropyridine ring in N-methyl-indenopyridine proceeds via acid hydrolysis of enamines.[3] The molecule of dihydroindenopyridine **2** contains two structural units which can be regarded as enamines and for that reason the fission of the N-C$_2$ and N-C$_{9b}$ bonds leads to two intermediates, A and B (*Scheme 3*), which could not be detected in the reaction mixture. However the sub-sequent products **10** and **11** could be isolated suggesting the cleavage of a strictly definite bond. The formation of the indenopyridone **9** as a result of intramolecular interaction of the ester group with the NHCH$_3$ function of intermediate B confirms the cleavage of the N-C$_2$ bond in dihydroindenopyridine **2**. 2-Benzylideneindan-1,3-dione **10** can be formed from both intermediates *A* and *B* (by cleavage of the N-C$_{9b}$ or N-C$_2$ bonds), however experimental findings show that triketone **11** under similar reaction conditions does not follow the retro-Michael reaction.

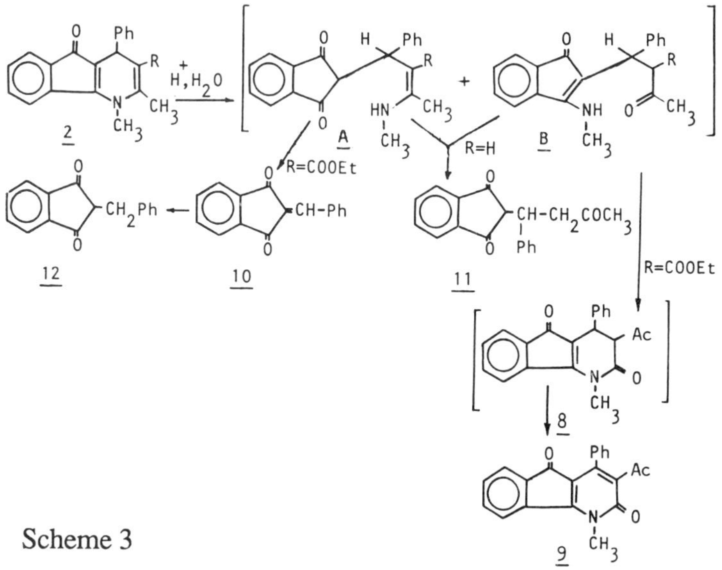

Scheme 3

Consequently, benzylideneindandione is formed directly from the intermediate enamine *A*, the rate of its cleavage as well as the rate of recyclization of intermediate *B* into pyridone **9** being higher than the rate of intermediate enamine hydrolysis. The process is determined by the ring-opening step of the starting indenopyridine. Intermediate *A* alone takes part in the formation of benzylideneindan-1,3-dione, whereas intermediate *B* undergoes the retro-Michael reaction to give indan-1,3-dione and 2-benzylidene derivative of acetoacetic ester. The benzylideneindandione is partially reduced to 2-benzylindan-1,3-dione **12** by the intermediate tetrahydroindenopyridone **8** as well as by the starting dihydroindenopyridine **2**. The ratio of the recovered hydrolysis products (**10** +**12**) : **9** = 4.1 serves as evidence that splitting of the N-C$_{9b}$ bond of dihydroindenopyridine **2** proceeds faster than the splitting of the N-C$_2$ bond.

The formation of the triketone **11** as a result of 1,2-dimethyl-4-phenyl-5-oxo-4,5-dihydroindeno[1,2-b]pyridine (**2**, R=H) hydrolysis is not indicative of the cleavage of a definite bond. The cleavage of the N-C_9 bond seems more likely, because in the absence of the 3-COOEt group a significant electronic shift to the C_{4a} atom is observed. Due to the absence of this group the intermediate product *B* also fails to undergo recyclization (*Scheme 3*).

The N-substituted 2-methyl-3-ethoxycarbonyl-4-phenyl-5-oxo-4,5-dihydroindeno[1,2-b]pyridine under the same acidic hydrolysis conditions does not undergo a hydrolytic ring-opening.

4H-4a,5-Dihydroindeno[1,2-b]pyridine

As mentioned above, 4$\underline{H}$,4a,5-dihydroindenopyridines (**3**) were obtained by alkylation of 1$\underline{H}$-4,5-dihydro isomers. In order to obtained by alkylation of 4a,5-dihydroindenopyridones (up to 60-65%), increase the alkylation must be performed in a protondonating medium, where the yield of N-alkyl derivatives **2** (by-products in this case) is less than 10% and does not interfere with the purification of 4a,5-dihydroindenopyridines.

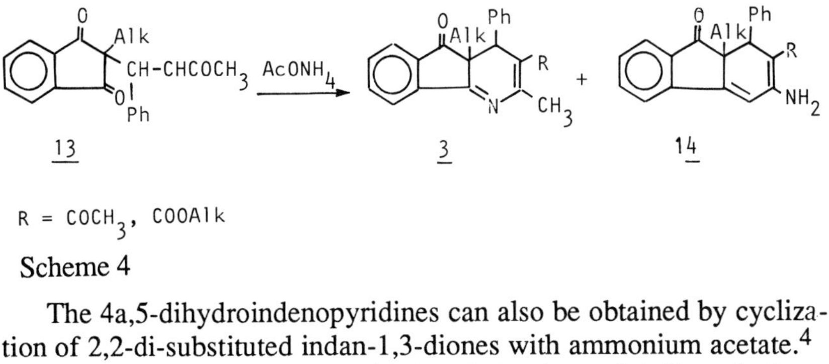

R = $COCH_3$, COOAlk

Scheme 4

The 4a,5-dihydroindenopyridines can also be obtained by cyclization of 2,2-di-substituted indan-1,3-diones with ammonium acetate.[4] The δ-dicarbonyl compounds, **13**, *i.e.* derivatives of 4-(2-methylindan-1,3-dion-2-yl)-4-phenylbutene-2-one, containing electron withdrawing substituents at C-3, undergo ring-closure (*Scheme 4*). The first step of the reaction is the formation of an imino derivative from the non-cyclic carbonyl group with subsequent ring-closure of the heterocyclic or carbocyclic system involving the carbonyl group in the indan-1,3-dione moiety. Interaction of the latter group with the amine function of the putative intermediate leads to the formation of dihydroindenopyridine **3**, interaction with the terminal methyl group leads to the formation of 3-aminofluorenone **14**. The ratio of products **3** and **14** depends on the reaction medium. The formation of indenopyridine **3** is significantly effected by the introduction of aprotic solvents into the reaction medium, but the formation of aminofluorenone is favoured by the presence of water (in dilute 50% acetic acid **3** is not formed).

It has been shown by X-ray analysis[5] that the dihydropyridine ring of 4a,5-dihydroindenopyridine **15** adopts a distorted boat-type conformation. The mean plane formed by the four atoms C_2-C_3-C_{4a}-C_{9b} is folded up along the line C_2-C_{4a} towards the N and C_4 atoms. The planes formed by the atoms N-C_2-C_{9b} and -C_3-C_4-C_{4a} depart from the mean plane (the angles are equal to 10° and 40°). The indene moiety of the molecule is almost planar and forms a 19° angle with the mean plane. The substituents at C-4 and C-4a are oriented diaxially.

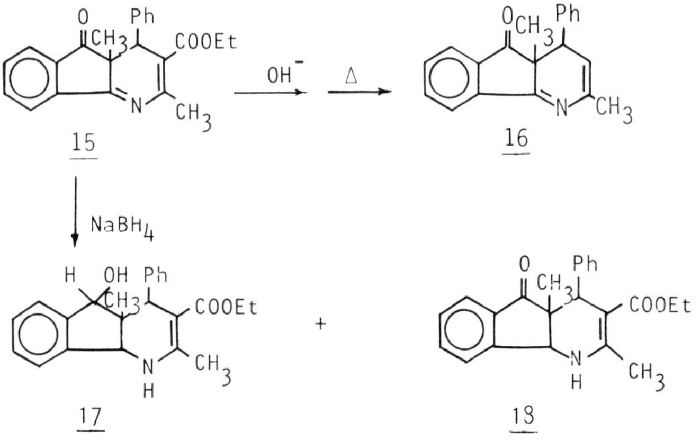

Scheme 5

The ester group of indenopyridine **15** can be easily hydrolysed. The 2,4a-dimethyl-4-phenyl-5-oxo-4H-4a,5-dihydroindeno[1,2-b]-pyridine 3-car-boxylic acid obtained undergoes thermal decarboxylation to yield the dihyd-roindenopyridine **16**, which can also be obtained by alkylation of compound (**1**, R=H). Reduction of the indenopyridine **15** with sodium boro-hydride gives a mixture of tetrahydroindenopyridines **17** and **18**, whereas catalytic hydrogenation of **15** leads to 1H-4,4a,5,9b-tetra-hydroindeno[1,2-b]-pyridine **18** in quantitative yield (*Scheme 5*).

On boiling 2,4a-dimethyl-3-ethoxycarbonyl-4-phenyl-5-oxo-4H-4a,5-di-hydroindeno-[1,2-b]pyridine **3** in acidic medium the N=C_{9b} bond undergoes hydrolysis[4] resulting in dihydropyridine ring cleavage to give triketone **13** and also 1,9a-dihydrofluorene-9-one derivatives **14** and **19** (*Scheme 6*). The first step of the ring-opening involves protonation of the nitrogen atom and addition of water to C_{9b} with subsequent fission of the single C-N bond. Hydrolysis of indeno-pyridine **3** yields compound **13** as the main product resulting from nucleophilic substitution of the amino group in the intermediate A. Recyclization of the primary hydrolysis product, i.e. intramolecular interaction of the terminal CH_3 group with the CO group in the indan-1,3-dione moiety producing compound **14** proceeds in parallel to substitution. 3-Hydroxyfluoren **19** does not result directly from hydro-lysis, but is formed upon intramolecular condensation of **13**, as supported by experimental evidence.

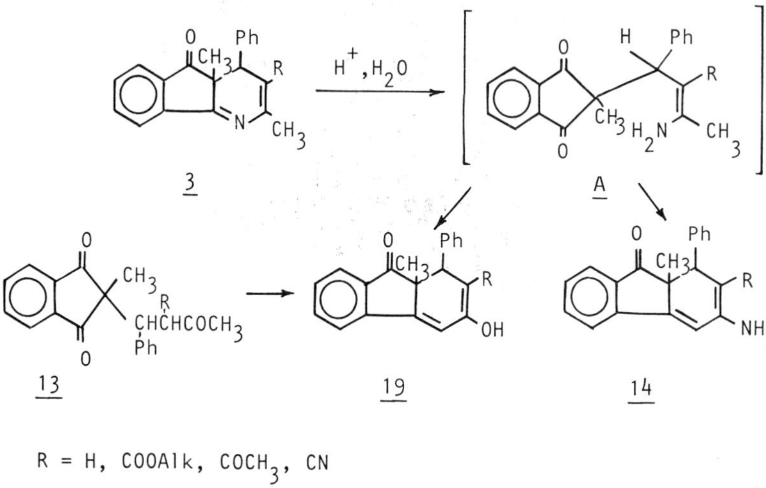

R = H, COOAlk, $COCH_3$, CN

Scheme 6

5,9-Dihydroindeno[1,2-b]pyridines

The transformation of 1,4-dihydropyridines to the appropriate 1,2-dihydroisomers in a series of monocyclic pyridines was carried out by reduction of pyridinium salts. The required indenopyridium salts **21** were obtained[6] in two ways: by alkylation of indenopyridines **20** or by oxidation of their N-alkyl-4,5-dihydroderivatives **2** (*Scheme 7*). The method previously used for the synthesis of pyridinium salts, involving oxidation of N-methyl-1,4-dihydro-pyridines by hydrogen peroxide in the presence of perchloric acid in the case of 4,5-dihydroindeno-pyridines **2** is complicated by cleavage of the dihydropyridine ring in acidic medium.

Reduction of indenopyridinium perchlorates **21** by sodium borohydride proceeds selectively to give 5,9b-dihydroindenopyridines **22**. The treatment of 1,2-dimethyl-3-ethoxycarbonyl-4-phenyl-5-oxo-5,-9b-dihydroindeno[1,2-b]pyridine with acid leads to isomerisation and ring-cleavage of the dihydropyridine ring and results in the formation of indenopyridone **9** and 2,5-dihydroindenopyridine **23** (*Scheme 7*). The formation of indenopyridone **9** is believed to occur according to the recyclization mechanism described previously and the formation of 1,2-dimethyl-3-ethoxycarbonyl-4-phenyl-5-oxo-2,5-dihydroindeno-[1,2-b]pyridine **23** appears to result from sigmatropic rearrangement.

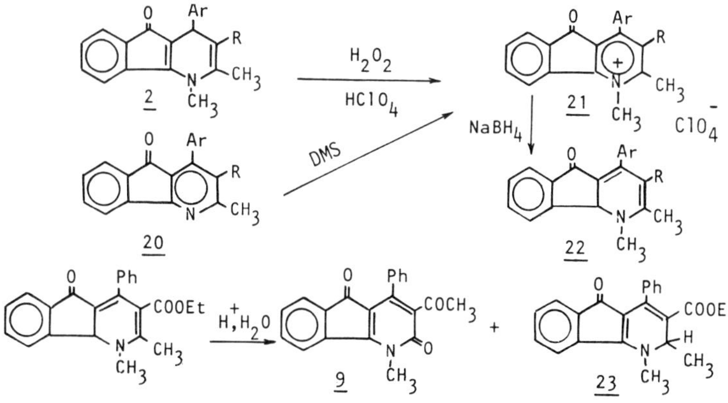

Scheme 7

Tetrahydroindeno[1,2-b]pyridines

Treatment of 4a,5-dihydroindeno[1,2-b]pyridines **3** with sodium boro-hydride gives a mixture of products with a reduced N=C bond and a 5-CO- group. Catalytic hydrogenation of 4$\underline{H}$-4a,5-dihydroindenopyridines **3** is a convenient method for the preparation of 4,4a,5,9b-tetrahydroindenopyridines **24** in quantitative yield. The hydrogenation of the N=C bond is stereoselective leading exclusively to one stereoisomeric form of tetrahydroindenopyridine (*Scheme 8*).

1$\underline{H}$-Tetrahydroindenopyridines **24** undergo alkylation with methyl iodide in aprotic medium in the presence of an alkaline agent to produce N-methyl derivatives **25**. Interaction of 1$\underline{H}$-tetrahydroindenopyridines **24** with alkylation agents capable of stable carbonium ion formation, e.g. with CH_3O-CH_2Cl or benzhydryl iodide, gives 4H-4a,5-dihydroindenopyridines **3** instead the expected alkylation products, thereby, the carbocation acting as hydride ion acceptor during the oxidation of the tetrahydroindenopyridine.

R = CN, $COCH_3$, COOAlk
R' = CH_3, OEt

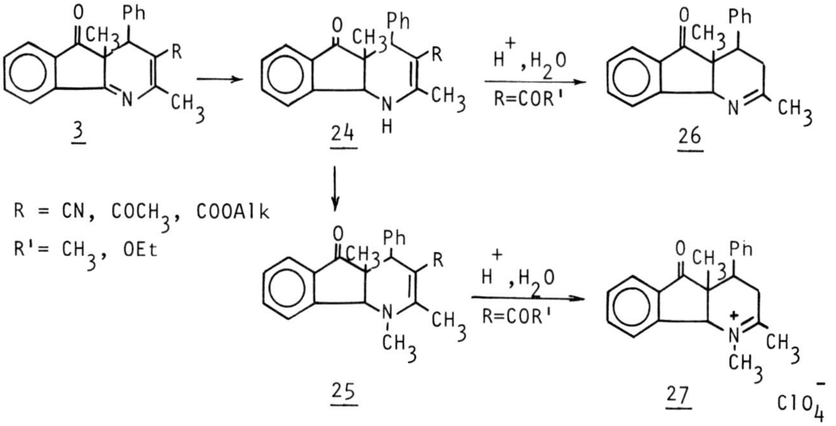

Scheme 8

In acidic medium, isomerisation of tetrahydroindenopyridines **24** and **25** containing a carbonyl function at C-3 takes place. This process is accompanied by elimination of the carbonyl function. Isomerisation of N-unsubstituted 1H-tetrahydroindenopyridines **24** results in 3H-4,-4,a,9b-tetrahydroindenopyridines **26**, whereas N-methylderivatives **25** with acid form 1-methyl-3H-4,4a,5,9b-tetrahydroindenopyridin-ium cations isolated as perchlorates **27**. In contrast to the above compounds, 3-cyano derivatives of (**3**, R=CN) remain unchanged under these conditions and can be recovered quantitatively from the reaction mixture.

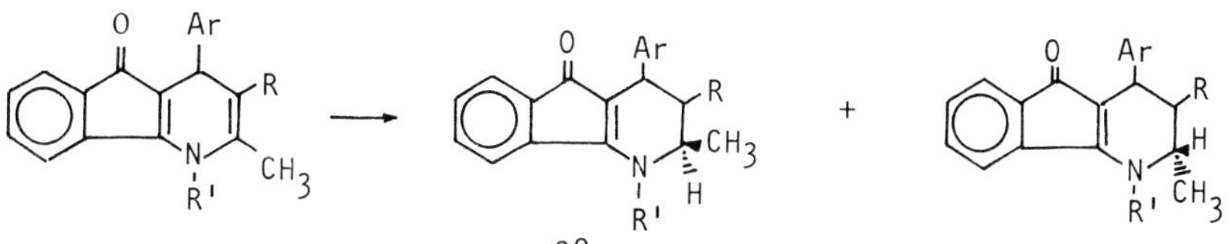

R = COOEt, COCH$_3$, CN
R' = H, CH$_3$

Scheme 9

Reduction of the C$_2$=C$_3$ bond in 1H-4,5-dihydroindenopyridines and their N-methyl derivatives with triethylsilane in trifluoroacetic acid gives a mixture of diastereomeric 2,3,4,5-tetrahydroindenopyridines **28** and **29** with different configurations at the C$_2$ atom (*Scheme 9*).

REFERENCES

1. V. Lusis, D. Muceniece, A. Zandersons, I. Mazeika, G. Duburs, Khimia Geterotsikl. Soedin.,1984, 393.
2. V. Lusis, D. Muceniece, G. Duburs, Khimia Geterotsikl. Soedin., 1986, 709.
3. V. Lusis, D. Muceniece, G. Duburs, Khimia Geterotsikl. Soedin., 1986, 1363.
4. V. Lusis, D. Muceniece, G. Duburs, Tetrahedron, 1986, **42**, 1547.
5. V. Lusis, D. Muceniece, A. Zandersons, I. Mazeilka, G. Duburs, Khimia Geterotsikl. Soedin.,1984, 1388.
6. A.Zandersons, V. Lusis, D. Muceniece, G. Duburs, Khimia Geterotsikl. Soedin., 1986, 88.
7. D. Muceniece, A. Zandersons, V. Lusis, G. Duburs, Khimia Geterotsikl. Soedin., 1987, 86.

Synthetic Studies on Biologically Active Isoquinoline Alkaloids

L. Castedo, D. Domínguez, and E. Guitián

DEPARTAMENTO DE QUÍMICA ORGÁNICA (FACULTAD DE QUÍMICA) Y SECCIÓN DE ALCALOIDES DEL CSIC, UNIVERSIDAD DE SANTIAGO DE COMPOSTELA, 15706 SANTIAGO DE COMPOSTELA, SPAIN

1 INTRODUCTION

We have been interested during the past few years in the synthesis of isoquinoline alkaloids, particularly aporphines (**1**) and the oxidized members of this group, dehydroaporphines (**2**), oxoaporphines (**3**), quaternary oxoaporphines such as corunnine (**4**), 4,5-dioxoaporphines (**5**) and aristolactams (**6**).[1]

From the synthetic point of view, an important feature of the aporphinoids is the presence of a phenanthrene or a dihydrophenanthrene system (rings A, C and D) in the molecule. The classical methods of preparing aporphinoids (**7**) are based on the biogenetic route, which means considering the biaryl bond "a" as the strategic one. Formation of this bond can be achieved by a) phenolic or nonphenolic oxidative coupling, b) the Pschorr cyclization, or c) photocyclization of a stilbene or dihydrostilbene intermediate (**8**) (Scheme 1).

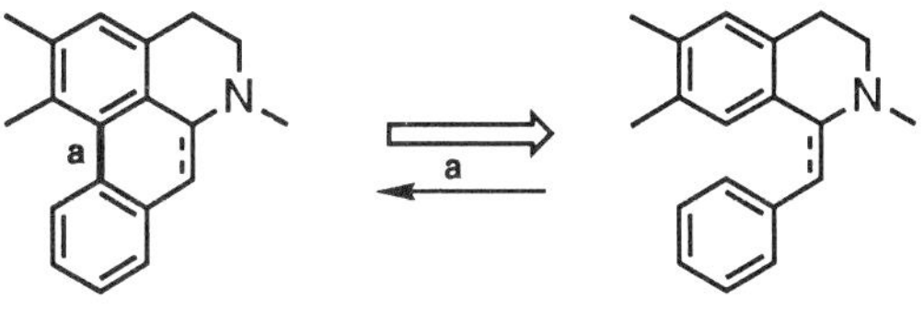

Biologically Active Isoquinoline Alkaloids

Scheme 1

As the classical methods are in some cases very laborious, with too many steps, and since they give unsatisfactory results when there are steric or geometrical impediments to cyclization, we decided to develop new methods of aporphinoid synthesis that combined generality, convergence, simplicity and efficiency. The result is a new synthetic procedure that is applicable to a variety of different kinds of isoquinoline alkaloid and of derivatives closely related to Amaryllidaceae and indole alkaloids.

Retrosynthetic analysis of the aporphine skeleton (**9**) led us to examine the possibility of simultaneously constructing bonds "a" and "b", or "a" and "c", by means of a Diels-Alder reaction in the latter stages of the synthesis (Scheme 2); that is, by an intermolecular reaction between a methyleneisoquinoline (**10**) and an aryne (**11**) or by intramolecular reaction between a styrene and an aryne (**12**). We have named the former the **I**ntermolecular **B**enzyne **C**ycloaddition (IBC) approach.

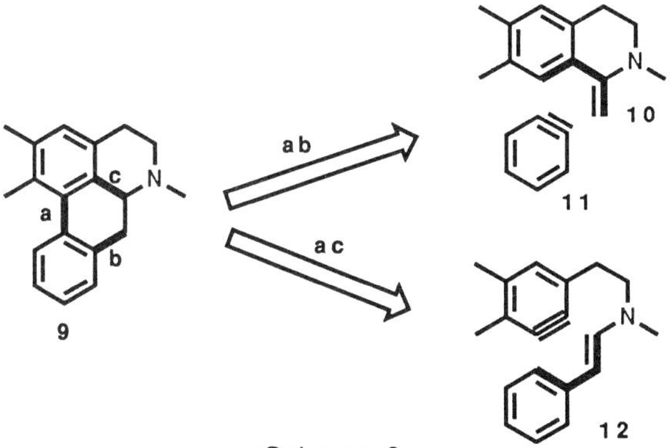

Scheme 2

2 INTERMOLECULAR BENZYNE CYCLOADDITION

In pursuing the IBC approach we found that slow addition of an excess of both anthranilic acid and isoamylnitrite in DME to a refluxing DME solution of **13** afforded a 40% yield of dehydroaporphine **15**.[2] This transformation apparently involves dehydrogenation of the initially formed (4+2) cycloadduct **14** by the benzyne (**11**) (Scheme 3). Surprisingly, the dihydroderivative of **15** could not be detected in the reaction mixture. Note that the nitrogen atom is protected as an amide to prevent its being attacked by the benzyne.

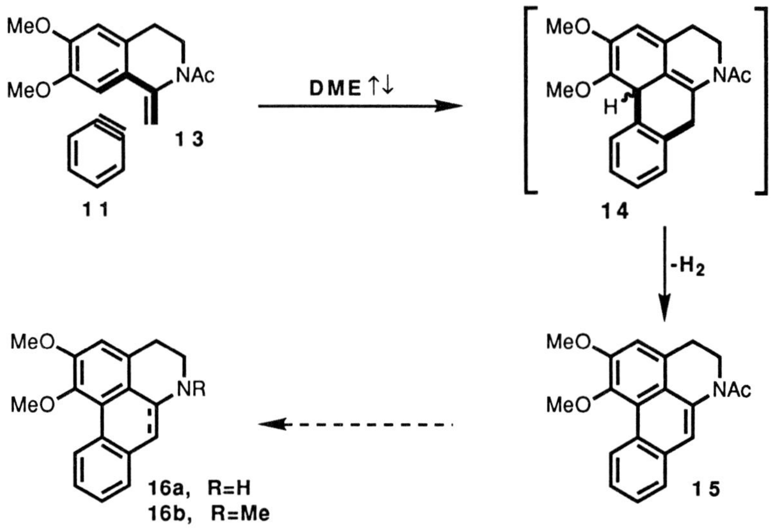

Scheme 3

Application of this IBC approach to the synthesis of noraporphines (**16a**) and aporphines (**16b**) requires the nitrogen atom to be protected with an easily removable group. We therefore, decided to investigate the reactivity of benzyne with various kinds of methyleneisoquinolines, with the nitrogen atom protected by various protecting groups.

The methyleneisoquinolines (**19-23**) (with two methoxy groups or one methylenedioxy group) are readily obtainable from the corresponding dihydroisoquinoline (**18**), which in turn can easily be prepared by acetylation of compound **17a** or **17b** with acetic anhydride/pyridine followed by Bischler-Napieralski cyclization (Scheme 4). Subsequent treatment of **18** with various acylating agents afforded the desired methyleneisoquinolines (**19-22**) in 70 to 90% yield,[3-7] depending on the compound (except in the case of **23b**, which was obtained in only 35% yield due to its ready hydrolysis[7]).

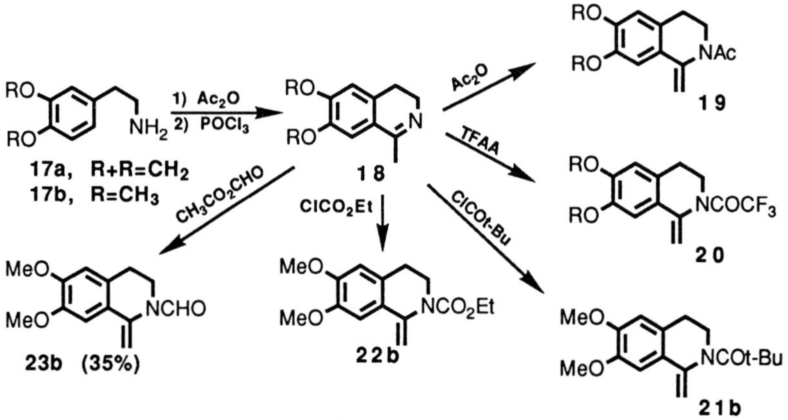

Scheme 4

We then found that compounds **19-22** reacted with benzyne, which was obtained by the preformed salt method,[8] to give the adducts **24** in yields of about 50 to 60%[6,7,9] (Scheme 5). We thus proved that acetyl, trifluoroacetyl, pivaloyl and ethoxycarbonyl are effective protecting groups for the nitrogen atom. However, the N-formyl derivative **23b** produced a complex mixture of products containing such a small amount of the desired (4+2) adduct that it could only be detected by NMR.[7]

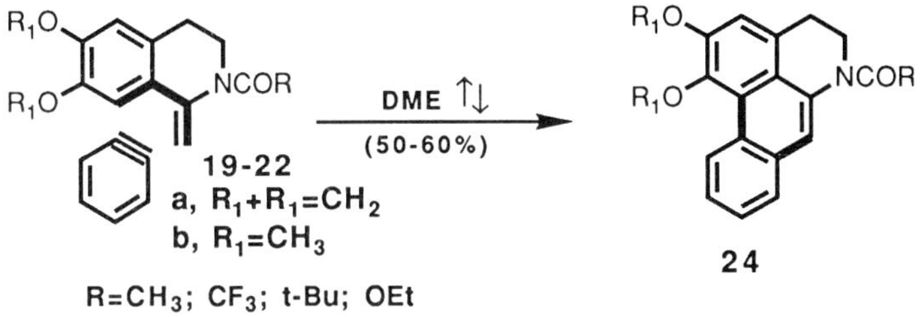

Scheme 5

The above (4+2) cycloaddition reaction is also observed when the aromatic ring of the styrene has no OR substituents[10] (Scheme 6). For example, methyleneisoquinoline **25** reacts with benzyne to give the dehydroaporphine **26** in 48% yield. Styrene **27** produces a mixture of **28** and **29** in 26% and 7% yield respectively, the latter compound presumably being formed by an "ene" reaction.

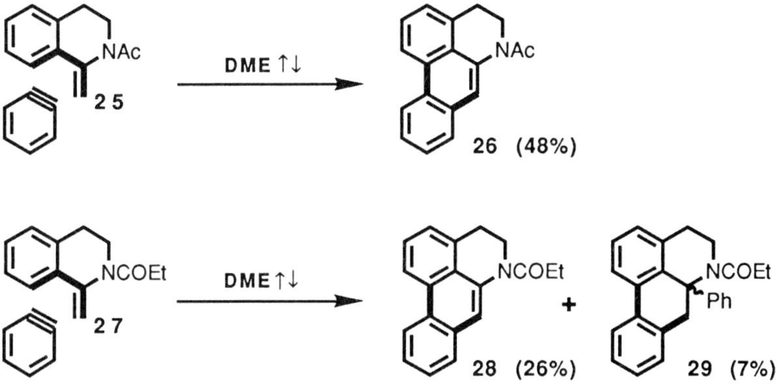

Scheme 6

The (4+2) adducts can be readily transformed into different aporphinoids (aporphines, noraporphines, 7-oxoaporphines, C7-substituted dehydroaporphines, etc.). For example, treatment of the trifluoroacetylderivative **30** with sodium borohydride afforded a 95% yield of the dehydronoraporphine **31**. Subsequent refluxing of **31** with aqueous formaldehyde in dioxane for two hours yielded the C7-substituted alkaloid duguenaine (**32**) in 82% yield (Scheme 7).

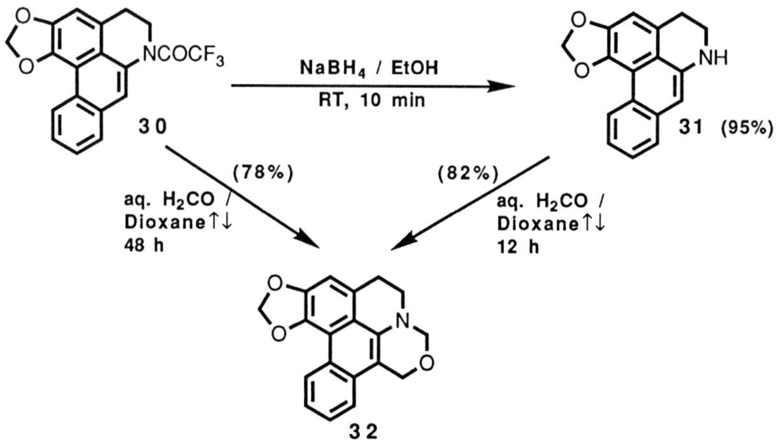

Scheme 7

Conversion of the N-trifluoroacetyl derivative **30** into duguenaine (**32**) was also achieved by simply refluxing compound **30** with aqueous formaldehyde as above, though for a longer time (48 hours).[7]

Application of the IBC approach to the synthesis of 4,5-dioxoaporphines (**33**) and aristolactams (**35**) requires preparation of amides **34** and **36** respectively (Scheme 8).

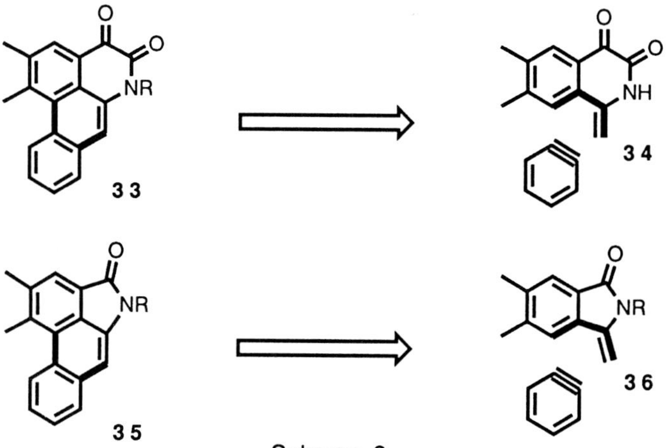

Scheme 8

In order to explore these possibilities, we first prepared the methylene derivative **40** by the following sequence of reactions, whose first step is the photooxidation of the amide **37** (Scheme 9). We assume that this gives the endoperoxide **38**, which in the presence of NaOMe affords the dione **39**. Under the acid conditions used to generate the aryne by aprotic diazotation, the methylene derivative **40** is formed; this reacts with benzyne to give norcepharadione B (**41**)[2] in 60% yield.[6]

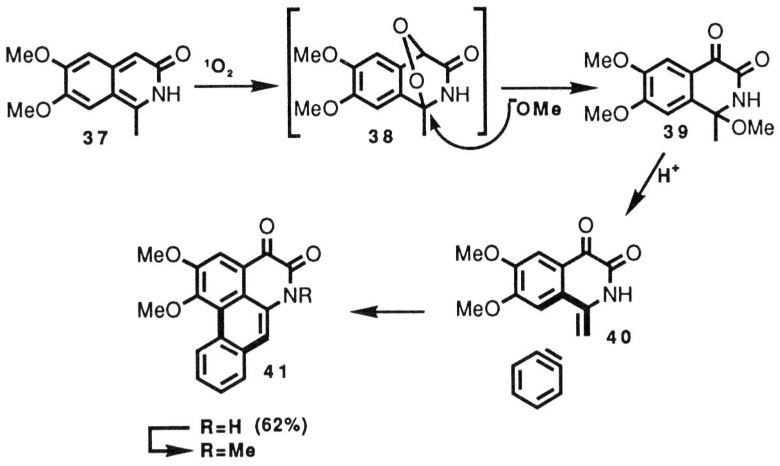

Biologically Active Isoquinoline Alkaloids

Scheme 9

Results obtained in the absence of substitution show that aristolactams (**43**) can also be obtained by the IBC approach,[2,6] though the yields are lower due to geometrical problems (Scheme 10). The styrene compound **42** reacts with benzyne to give a 27% yield of aristolactam **43** and a 63% yield of benzocyclobutene **44**, which results from an unconcerted (2+2) cycloaddition. The increased distance between positions 1 and 4 of the diene system appears to hinder the Diels-Alder reaction in this case.

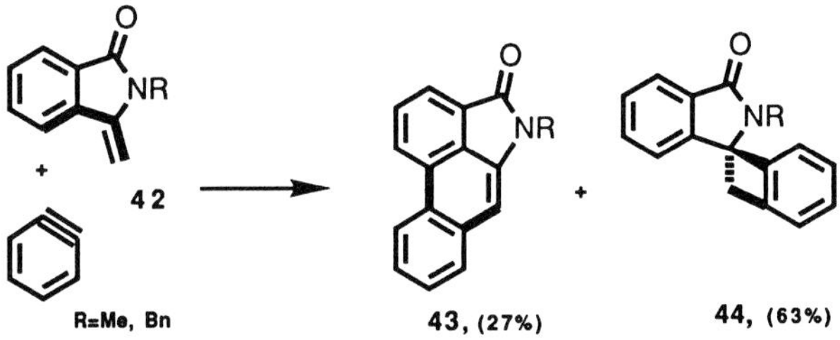

Scheme 10

Having achieved the above results, our next step was to investigate the viability of the IBC approach for preparing aporphinoids with substituents on ring D. The simplest case is that of symmetrically substituted arynes such as **45** (Scheme 11). We found that **45** reacted with the methylene derivative **40** in the desired way to give norpontevedrine (**46**),[11] though with a lower yield (20%) than in the case of unsubstituted benzyne. Norpontevedrine (**46**) can readily be transformed into the alkaloid pontevedrine (**47**) by N-methylation with methyl fluorosulphonate and NaH.[12]

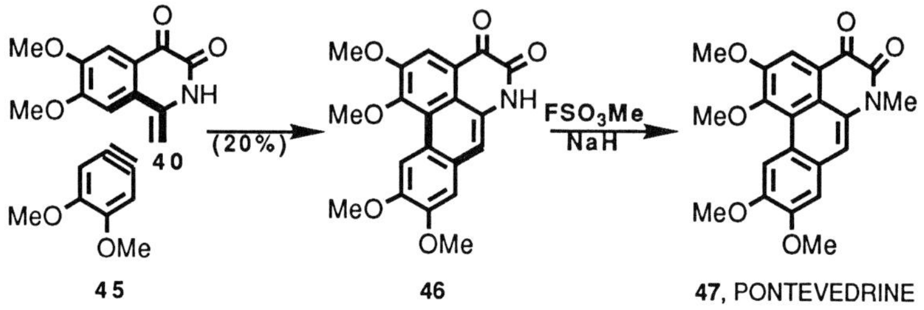

Scheme 11

When the aryne is asymmetrically substituted, regioselectivity is ensured by the presence of a methoxyl <u>ortho</u> to the aryne triple bond (whereas a mixture of isomers is produced if the methoxyl group is <u>meta</u> to the triple bond).

Methyleneisoquinoline **13** reacts with 3-methoxybenzyne (**48a**) and 3,4-dimethoxybenzyne (**48b**) to give a single isomer (**49**),[13] but reaction of compound **50** with 4-methoxybenzyne (**51**) produces a 1:1 mixture of regioisomers **52** and **53**[6] (Scheme 12). The regioselective reaction allows effective preparation of aporphinoids with substituents at both position 1 and position 11, which are difficult to obtain by the classical methods.

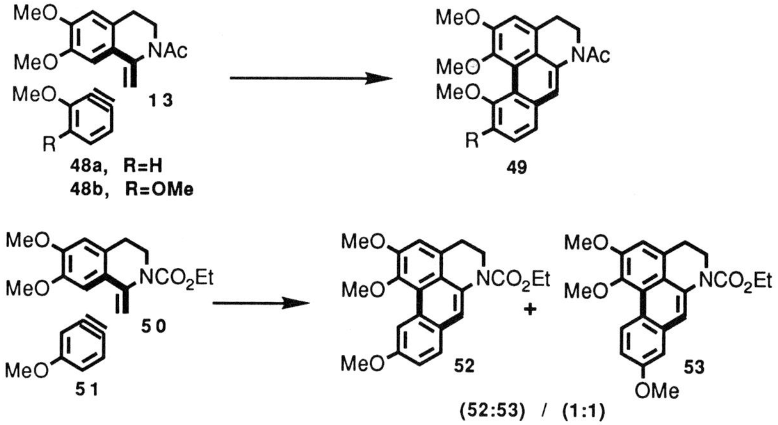

Scheme 12

Regioselectivity is also observed under these conditions if the aromatic ring of the starting styrene has no substituents, as is shown by the 34% yield of **54** (apomorphine substitution) obtained by reaction of compound **25** with 3,4-dimethoxybenzyne (**48b**)[10] (Scheme 13). Adduct **54** can readily be transformed in apomorphine (**55**) by conventional procedures (amide hydrolysis, reduction to dihydrophenanthrene, <u>N</u>-methylation and ether cleavage). We believe that this convergent, regioselective approach could be useful for the synthesis of apomorphine analogues of great pharmacological interest.

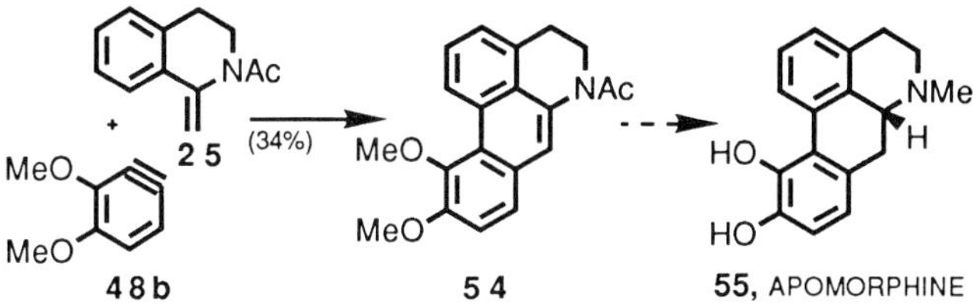

Scheme 13

The above methods, together with the easy transformation of dehydronoraporphines into oxoaporphines by oxidation with Fremy's salt, permitted simple synthesis of the quaternary oxoaporphine alkaloid PO-3 (**61**)[9] (Scheme 14). The structure of PO-3 had only been established on the basis of spectroscopic evidence, and it is difficult to obtain by the classical methods because of steric hindrance to cyclization. But the methylene compound **56** reacted regioselectively with 3-methoxybenzyne (**48a**) to give a 26% yield of dehydroaporphine **57**, which was transformed into the final product **61** by N-deprotection with sodium borohydride, oxidation of the resulting dehydronoraporphine (**58**) with Fremy's salt to oxoaporphine (**59**), N-alkylation with methyl iodide to **60**, and thermolysis.

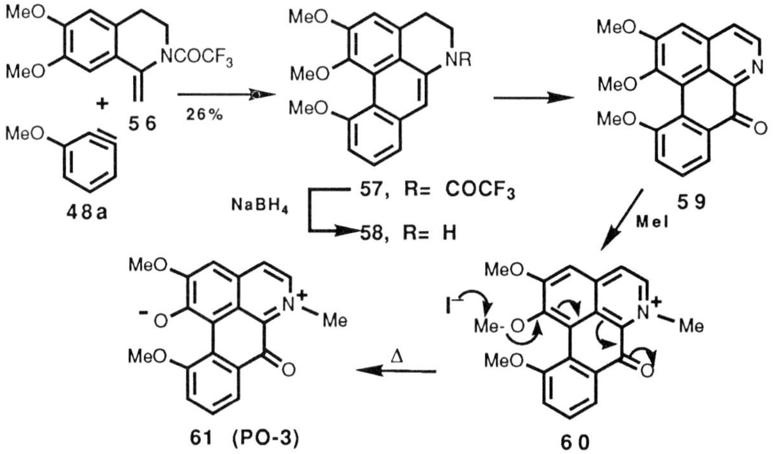

Scheme 14

The regioselectivity observed in the above cycloadditions can be explained in terms of an asynchronic concerted mechanism involving the transition state **66**, in which the formation of bond "c" is more advanced than that of bond "a" and the aryne has acquired a negative charge. However, the experimental findings are also compatible with a non-concerted mechanism involving the intermediate **64**, in which the formation of bond "c" and the anion result from the aryne being attacked by the more nucleophilic site of the diene system (Scheme 15).

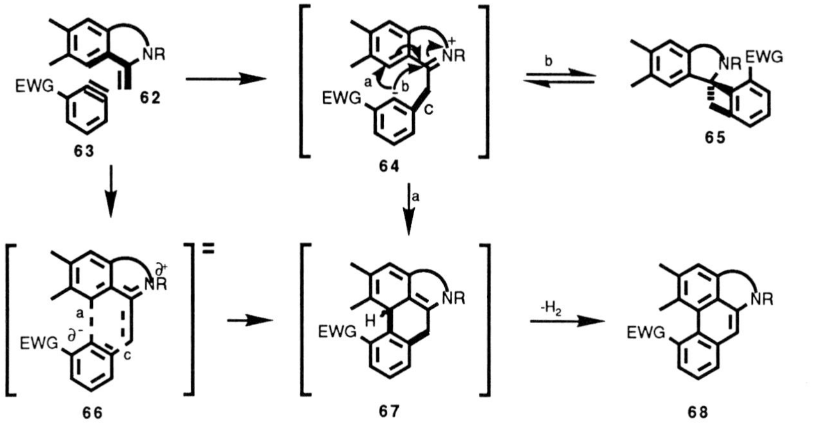

Scheme 15

On both hypotheses, the observed regioselectivity is due to the stabilization of the negative charge in **64** and **66** by the polarizing effects of the substituents, which are strong in the case of an ortho EWG such as a methoxyl, but very weak for more distant (meta and para) substituents. The intermediate **64** has two electrophilic sites; anion attack at one position should give the (4+2) cycloadduct **68** via the intermediate **67** and attack at the other position the (2+2) adduct **65**.

Our next goal was to investigate the reactivity of benzyne with methyleneisoquinolines that are substituted at the terminal position of the methylene group, which it was hoped would allow synthesis of C7-substituted aporphinoids. These include structure **70**, which has a methyl substituent at C7 and has recently been ascribed to the alkaloid trichoguattine on the basis of spectroscopic evidence. Surprisingly, reaction of benzyne with the monosubstituted methylene isoquinoline **69**, which is N-protected with a labile group, produced only a small amount of the expected C7 dehydroaporphine **70**, the major product being the dibenzoindolizine **71**[14,15] (Scheme 16). The structure of **70** was confirmed by independent synthesis along the lines of the classical photochemical approach, but its spectroscopic properties were different from those of trichoguattine. In conclusion, trichoguattine does not have the structure **70** originally assigned to it.

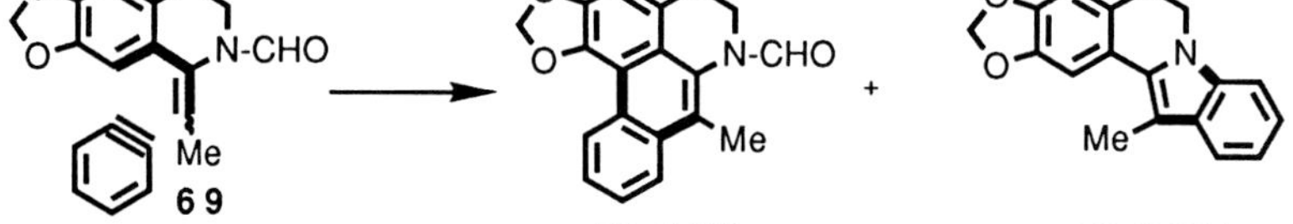

Scheme 16

A similar result was obtained when the monosubstituted methyleneisoquinoline **72** was reacted with benzyne[15] (Scheme 17). The structure of the formal (3+2)cycloadduct **74** was further confirmed by independent synthesis from compound **75**, which by photochemical electrocyclization produced **74** in 11% yield.

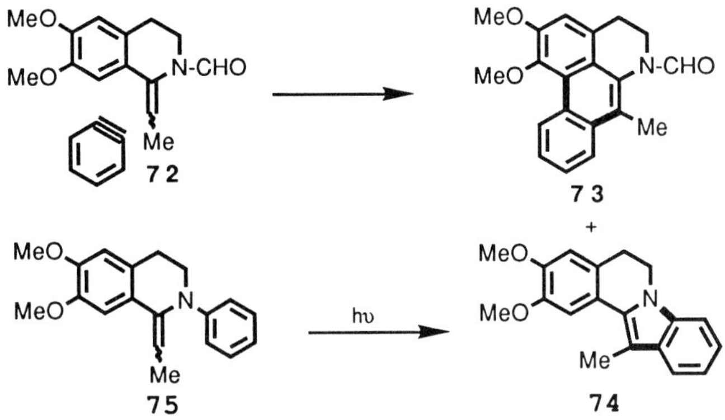

Scheme 17

The puzzling question is how the unexpected compounds such as **71** and **74** arise from the IBC approach. As a working hypothesis we thought that residual water and trichloroacetic acid in diazonium preparations might lead to the N-formyl compound **72** undergoing hydrolytic loss of the formyl group to give the imine **76** (Scheme 18). We therefore decided to investigate the reactivity of imine **76** with benzyne, and found that it gives the (3+2) cycloadduct **74** in 38% yield. The formation of **74** from **76** can be explained assuming that attack by nitrogen on benzyne, followed by subsequent β hydrogen abstraction, might lead via **77** to intermediate **75**. This could then suffer thermal disrotatory 6π electrocyclization to an intermediate such as **78** capable of producing the final product (**74**) by way of benzyne-promoted dehydrogenation.

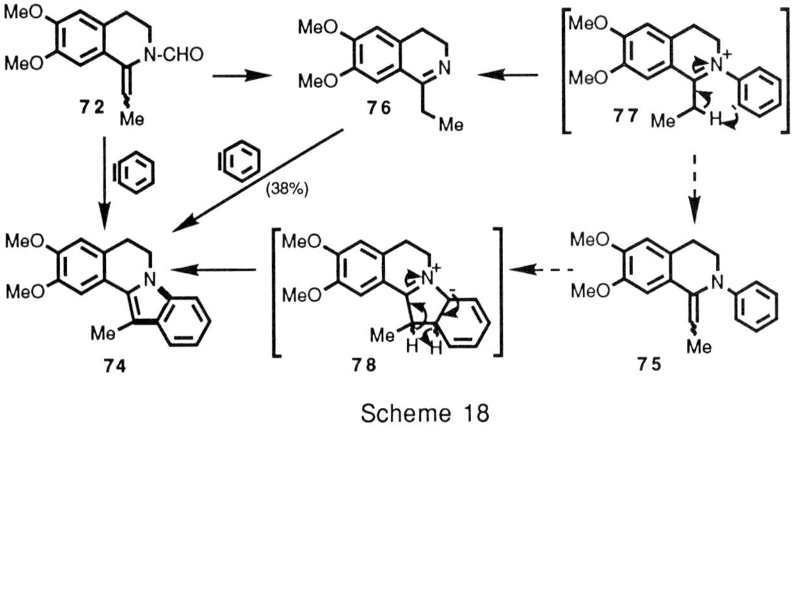

Scheme 18

In order to test this hypothesis we decided to study the reaction of the proposed intermediate **75** with benzyne under the same conditions (Scheme 19). We found that it did not give the desired (3+2) cycloadduct **74**, but instead the usual (4+2) cycloadduct **79** together with ketone **80**, a product possibly derived from the reaction of benzyne with the hydrolysis product of the enamine **75**.

The postulated mechanism must accordingly be rejected, neither **79** nor **80** having been detected in the reaction of **76** with benzyne. Further work is necessary to find the mechanism of this unexpected transformation.

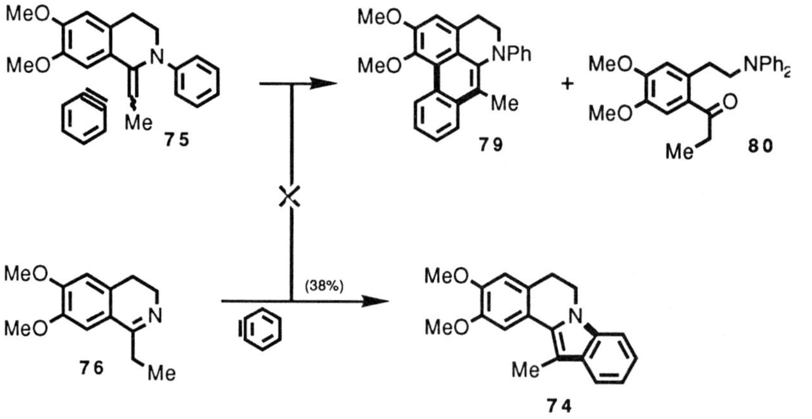

Scheme 19

Continuing with our work on the reactivity of monosubstituted methyleneisoquinolines with benzyne, we next studied compounds with less labile N-protecting groups such as carbethoxyl. We found that **Z-81** and **E-81** and a mixture of the two all gave a 50% yield of the expected C7 methyldehydroaporphine **83**[16] (Scheme 20). Assuming that the (4+2) cycloaddition takes place by a concerted mechanism, the above results clearly show that the dehydrogenation does not take place by benzyne-promoted 1,4-syn elimination of hydrogen from cycloadducts **82** and **84**, since in **84** the hydrogens are trans. Note that PMR spectra of samples taken at different times showed that no E-Z isomerization of **81** took place during its reaction with benzyne.

An alternative mechanism based on 1,2-dehydrogenation at C6a and C7 in **86**, a hypothetical intermediate supposed to derive from the initially formed cycloadducts **82** and **84** by a non-concerted 1,3 hydrogen migration, can also be ruled out, since N-acetylnorglaucine (**87**) did not yield the corresponding 6a,7-dehydroderivative under reaction conditions typical of the IBC approach, which led to **87** being recovered unchanged in high yield.

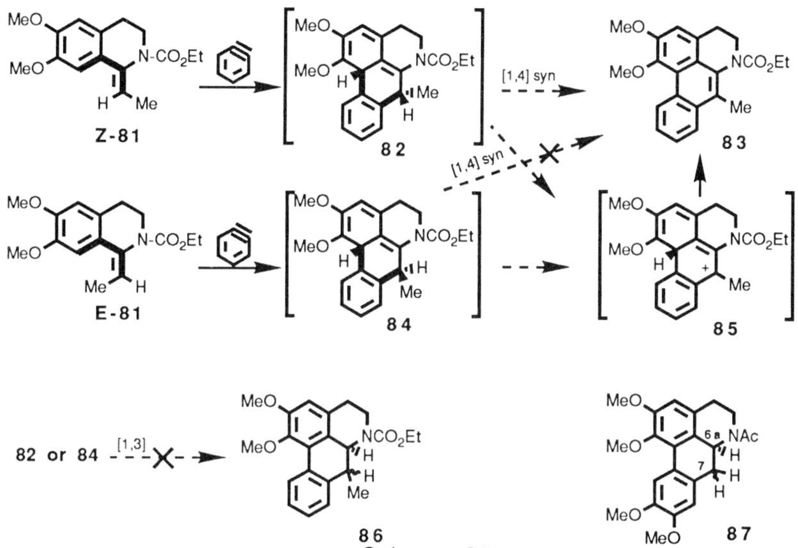

Scheme 20

Although other mechanisms cannot be ruled out, our findings are in keeping with that proposed by Heaney, in which the dehydrogenation takes place by benzyne-promoted loss of a hydride from the (4+2) cycloadduct **82** or **84**, followed by proton loss from the resulting carbenium (**85**).

Our next step was to study the reactivity of arynes with compounds such as **89**, containing a modified diene system in which the enamide is part of a pyrrolinedione heterocycle (Scheme 21). This type of compound is easy to prepare by treating an imine (**88**) with oxalyl chloride. Should the diene **89** react with an aryne in the normal way, it would give the isatine derivative **92**, which we knew from related studies to be transformable into naturally occurring oxoaporphines (**93**). Surprisingly, the reaction between the

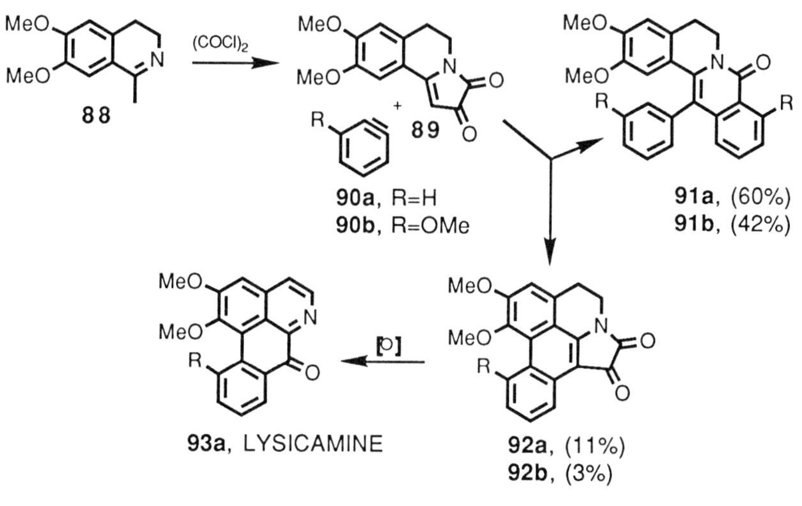

Scheme 21

benzyne **90a** and **89** produced only an 11% yield of the expected pyrrolinedione **92a** (which was readily transformed into the oxoaporphine lysicamine (**93a**) by Fremy oxidation); the major product of the reaction was the 13-phenyl-8-oxoprotoberberine **91a**, which was obtained in 60% yield. Similar results were obtained when compound **89** was treated with the unsymmetrically substituted 3-methoxybenzyne. Interestingly, only compounds **91b** (42%) and **92b** (3%) were found, thus proving the highly regioselective nature of the reaction.[17]

How can the formation of this formal (4+2) cycloadduct be explained? It may be hypothesized that the formation of compounds **92** and **91** takes place via a non-concerted mechanism in which the most nucleophilic position of **89a** attacks the aryne to give the anion **94**, which can then follow two competing pathways, (a) and (b) (Scheme 22). The former involves attack by the anion at the aromatic position to give the isatine **92**, while its attacking the amide carbonyl would give the intermediate **95** and thence, via decarbonylation, reaction with a second benzyne molecule and isomerization of **96**, the oxoprotoberberine **91**. Enamide **89b** was shown not to be an intermediate in the reaction by its being recovered unchanged after treatment with benzyne under typical reaction conditions.

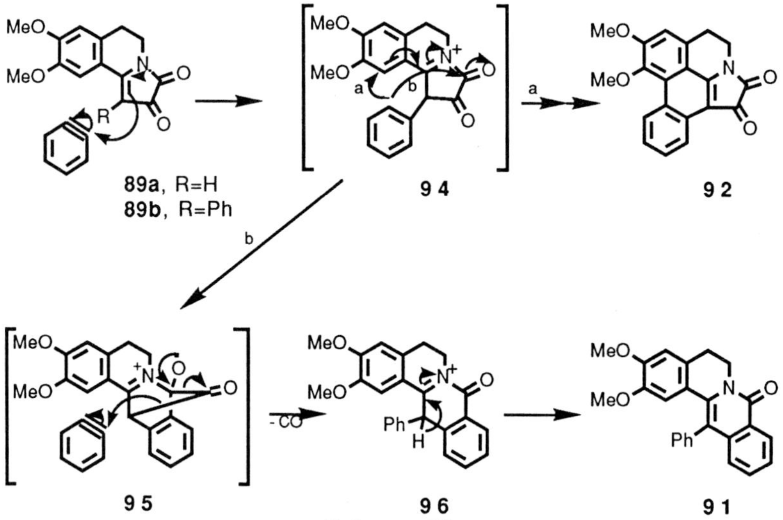

Scheme 22

The above behaviour of the pyrrolinedione system **89** opened the way to the synthesis of naturally occurring protoberberine alkaloids. Since in most natural protoberberines C13 has no substituent, or at most a methyl or a methoxyl group, our immediate aim was to avoid the arylation of C13 observed in **91**. Fortunately, this is achieved by the presence of a methyl at the corresponding position in the pyrrolinedione, which has enabled us to carry out total synthesis of corydaline (**101b**) (Scheme 23). The pyrrolinedione **97** reacts with benzyne (**98a**) to give the adduct **100a** in 51% yield, and with 3,4-dimethoxybenzyne (**98b**) to afford, regioselectively, a 32% yield of the adduct **100b**, which by reduction with LAH and sodium borohydride is converted into (±)-corydaline (**101b**).[17]

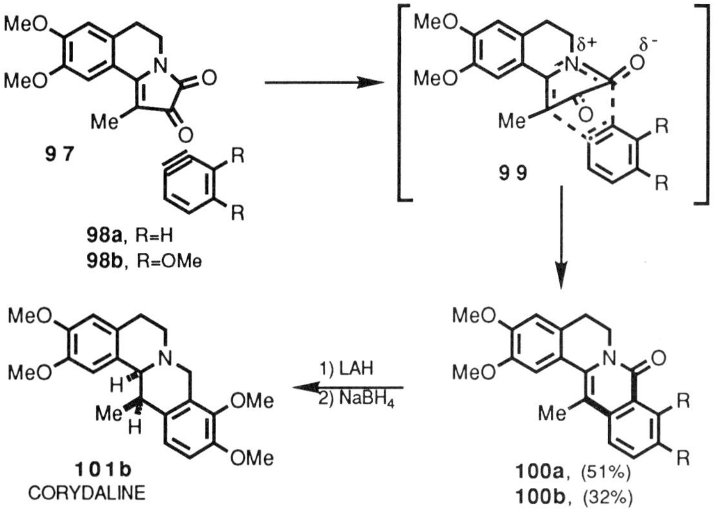
MeO
MeO
Me
97
98a, R=H
98b, R=OMe
99
1) LAH
2) NaBH4
101b
CORYDALINE
OMe
OMe
100a, (51%)
100b, (32%)

Scheme 23

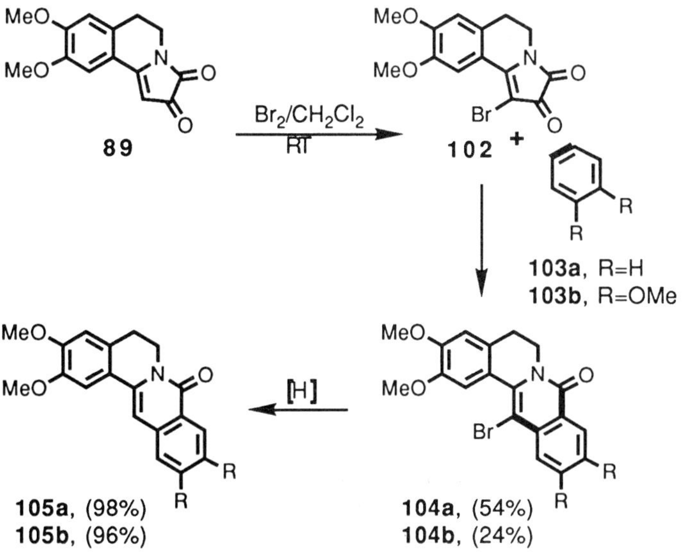
89
Br2/CH2Cl2
RT
102
Br
103a, R=H
103b, R=OMe
[H]
105a, (98%)
105b, (96%)
104a, (54%)
104b, (24%)

Scheme 24

Generalization of the above procedure requires that the group blocking phenylation of C13 be easy to remove once the protoberberine has been formed. Trials with various possible protecting groups have shown that bromine is effective in this role, as is illustrated by its use in the synthesis of oxyprotoberberine (**105a**) and of oxypseudopalmatine (**105b**) (Scheme 24), a minor ring-D-substituted alkaloid recently isolated from *Stephania suberosa.*

The bromopyrrolinedione **102**, which was readily obtained in 98% yield from pyrrolinedione **89** by direct bromination with bromine in dichloromethane at r.t., reacted with benzyne (**103a**) to give 13-bromo-8-oxyprotoberberine (**104a**) in 54% yield. This was debrominated in 98% yield to 8-oxyprotoberberine (**105a**) by hydrogenolysis. Similarly, reaction of bromoisatine **102** with 3,4-dimethoxybenzyne (**103b**) afforded a 24% yield of **104b**, which was transformed by hydrogenolysis into 8-oxypseudopalmatine (**105b**) in 96% yield.[18]

Having shown the potential of the pyrrolinedione approach for the synthesis of protoberberines, we have recently focused our attention on another important group of alkaloids, the indoles. We first investigated the viability of the pyrrolinedione strategy for preparing compounds with a β-carboline nucleus (**106**) from **107** (Scheme 25).

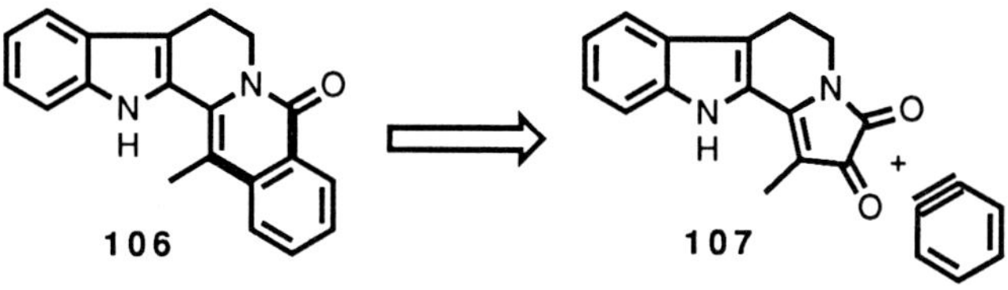

Scheme 25

We found that the pyrrolinedione **109**, which is readily prepared by treatment of imine **108** with oxalyl chloride, reacts with benzyne in refluxing DME to give a 49% yield of the desired (4+2) cycloadduct **110** (Scheme 26).[19]

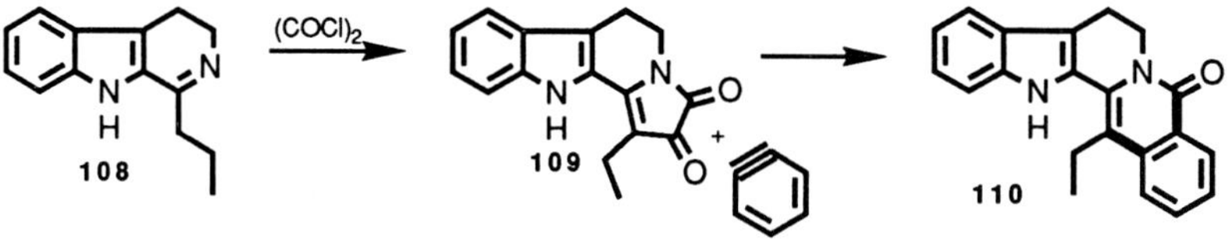

Scheme 26

We next studied the reactivity of pyrrolinediones with a removable group added at the unsubstituted ring D position to avoid its phenylation, as in the synthesis of protoberberines. We hoped in this way to be able to prepare unphenylated β-carboline derivatives, which by further manipulation of ring E might be convertible into β-carboline alkaloids (Scheme 27). Treatment of compound (**111**) with methyl oxalyl chloride in THF and pyridine (1 eq.), followed by reflux of the resulting esteramide in toluene in the presence of silica gel, afforded a 72% yield of isatine **112** This compound was readily brominated or chlorinated with bromine or N-chlorosuccinimide. We then found that the bromoisatine **113a** reacts with benzyne to give a mixture of compounds **114a** and **115a** in yields which ranged from 22 and 38% to 34 and 15%, depending on the excess of benzyne used. However, when the chloroisatine **113b** was reacted with three equivalents of benzyne, **114b** alone was obtained in 63% yield.[20]

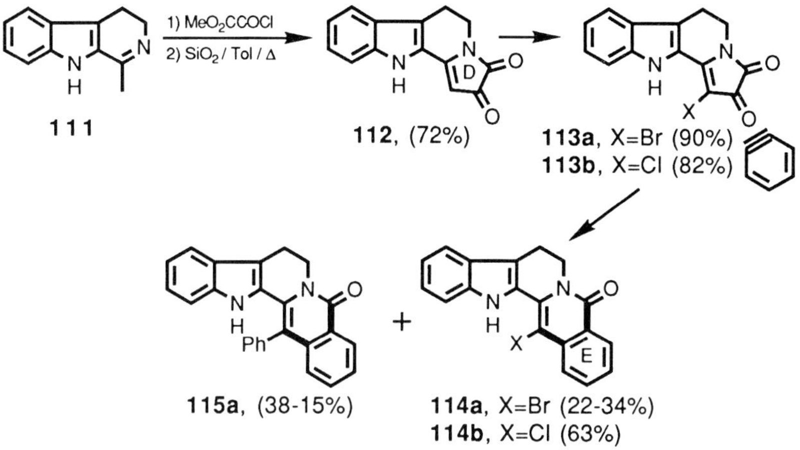

Scheme 27

These results show that in reactions with arynes, pyrrolinediones **116** act like azadienes **117**. One therefore wonders how they behave in reactions with other dienophiles. We have found, for example, that the pyrrolinedione **97** reacts rapidly with dimethyl acetylene dicarboxylate (DMAD) to give compound (**118**) with the tricyclic emetine skeleton[21] (Scheme 28).

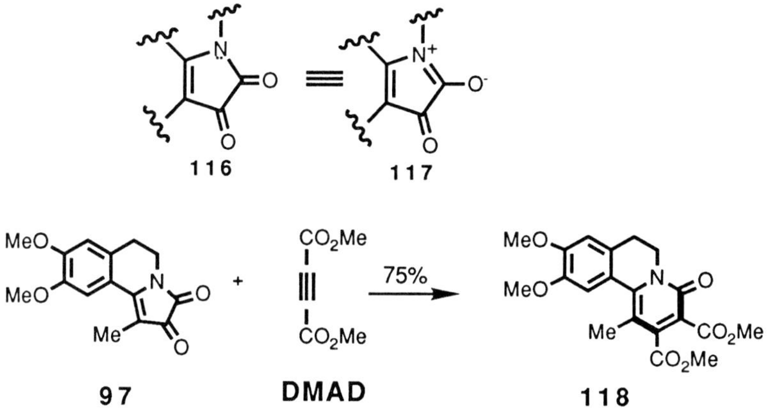

Scheme 28

More research is now being done to extend the potential of this pyrrolinedione approach to the synthesis of compounds related to various types of alkaloid such as emetine and corynantheine.

The pyrrolinedione method of synthesizing protoberberines and β-carbolines involves the formation of an isoquinolone that can be readily reduced to the corresponding isoquinoline. We reasoned that we ought to be able to take a similar route to other types of isoquinoline alkaloid such as the benzophenanthridines (**119**), among which there are a number of compounds of great pharmacological interest. Retrosynthetic analysis of the benzophenanthridine skeleton suggested that it should be possible to construct ring B of benzophenanthridines (**119**) by cycloaddition of an aryne (**122**) to the pyrrolinedione (**121**) (Scheme 29); this should give the amide **120**, which is readily transformable into **119**.

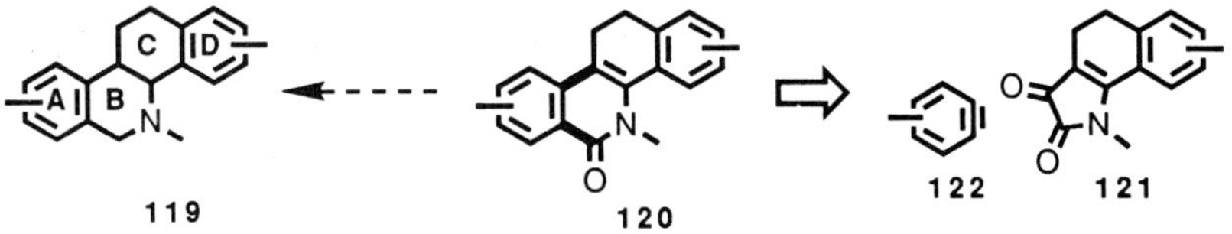

119 120 122 121

Scheme 29

In order to confirm this hypothesis, we synthesized the benzophenanthridines **125a**, **125b** and **125c** by reaction of benzyne with the pyrrolinediones **124a**, **124b**[22] and **124c**[23] (Scheme 30), which were readily obtained from the corresponding tetralones **123a**, **123b** and **123c** by reaction with methylamine and titanium chloride followed by oxalyl chloride.

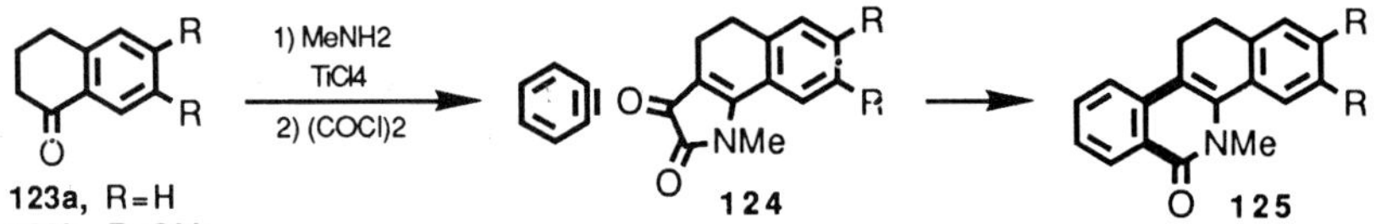

123a, R=H
123b, R=OMe
123c, R+R=OCH_2O

124 125

Scheme 30

In view of our success, we then used the pyrrolinedione procedure to prepare the ring-A-substituted benzophenanthridines **127a**, **127b** and **130**, the precursors of the alkaloids avicine (**128a**), nitidine (**123b**) and fagaronine (**131**) (Scheme 31); the latter two have considerable antitumor activity. Compounds **127a** and **127b** were obtained by reaction between pyrrolinedione **126** and arynes **45a** and **45b** respectively; and benzophenanthridine **130** by reaction of isatine **129** with aryne **45a.**[23]

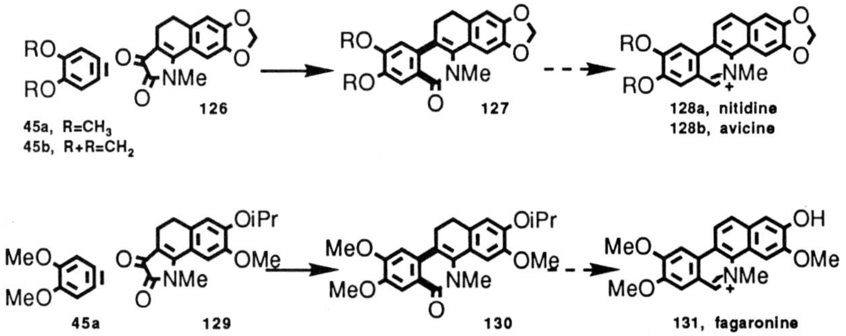

45a, $R=CH_3$
45b, $R+R=CH_2$

126 127

128a, nitidine
128b, avicine

45a 129 130 **131, fagaronine**

Scheme 31

Finally, we have also found that the pyrrolinedione **126** reacts regioselectively with the aryne **48b** to give the adduct **132**, which is the precursor of chelerythrine (**133**)[23] (Scheme 32).

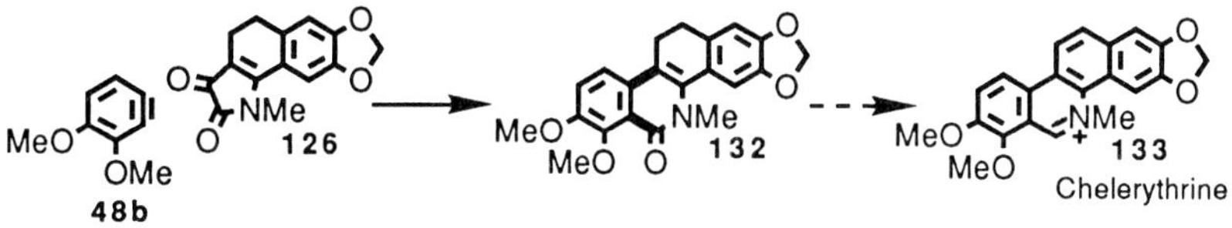

Scheme 32

As an alternative IBC approach to benzophenanthridines (**119**), we thought that it might be possible to synthesize these alkaloids constructing ring C by means of cycloaddition of an aryne (**122**) to a diene (**134**) (Scheme 33). When this strategy was attempted by Dyke using the diene **135**, only a 1.2% yield of the benzophenanthridine **136** was obtained.[24]

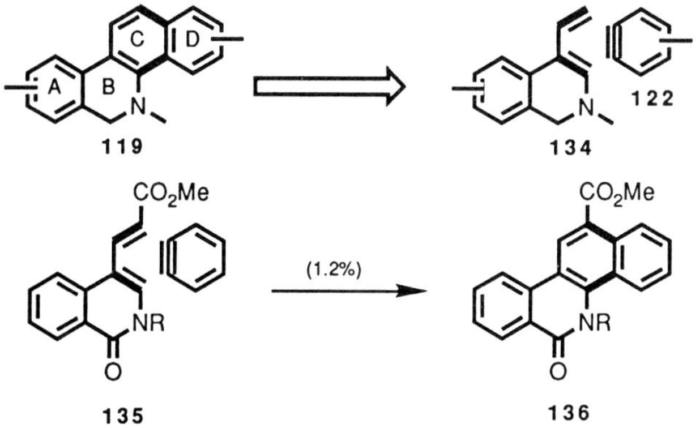

Scheme 33

We nevertheless felt that modification of the diene component might lead to better yields. After preliminary trials we chose the pyrone **137** as a suitable diene (Scheme 34). We found that treatment of this pyrone with an excess of preformed benzyne in refluxing DME/dioxane afforded a 43% yield of benzophenanthridine **140**, in which two benzyne molecules were incorporated, one by N-phenylation and the other by a Diels-Alder/retro Diels-Alder sequence, involving intermediate **138**.

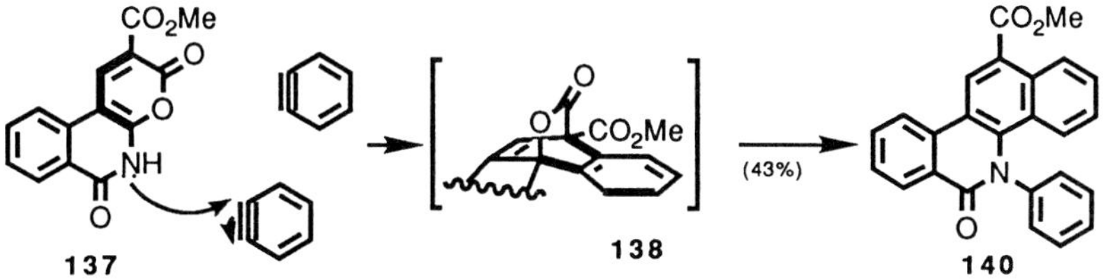

Scheme 34

To avoid the N-phenylation, pyrone **142**, in which a methyl group protects the nitrogen atom, was prepared as follows (Scheme 35). Treatment of the homophthalimide **140** with methyl orthoformate and acetic anhydride afforded a 94% yield of **141**, which by reaction with ethyl cyanoacetate and sodium methoxide gave a 94% yield of pyrone **142**. When this compound was treated for two hours with 20 equivalents of benzyne in refluxing DME, the only product was the adduct **144**, which has two benzyne molecules incorporated; but under milder conditions (5 equivalents of benzyne and 1 hour's reflux) we obtained both **143** and **144** (in 39% and 22% yields respectively), together with a 33% yield of the initial pyrone.[25] We eventually found that when the reaction is carried out for two hours with 15 equivalents of benzyne in refluxing dioxane, a 77% yield of **143** is obtained together with 15% of starting pyrone **142**.[26]

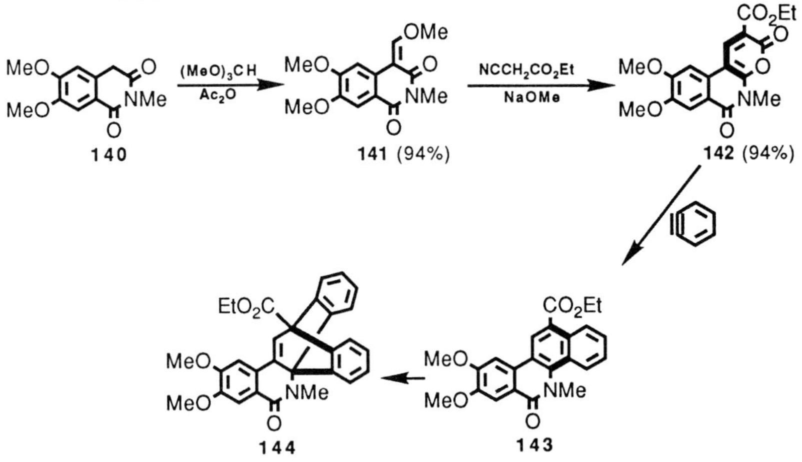

Scheme 35

3 INTRAMOLECULAR BENZYNE CYCLOADDITION

The second aryne strategy mentioned in Section 1 was the simultaneous formation of two bonds by intramolecular aryne cycloaddition between a diene and an aryne in the later stages of synthesis. The synthesis of aporphinoids (**145**), for example, might be achieved by forming bonds "a" and "c" from the precursor **146** (Scheme 36).

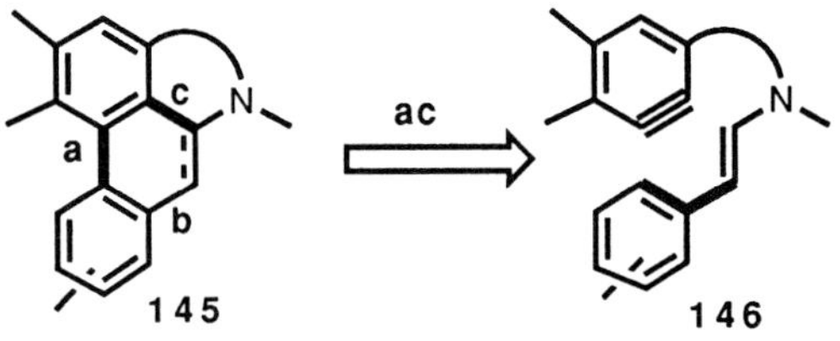

Scheme 36

We found that aristolactams can be synthesized by this intramolecular aryne approach as shown in the schemes 37 and 38.

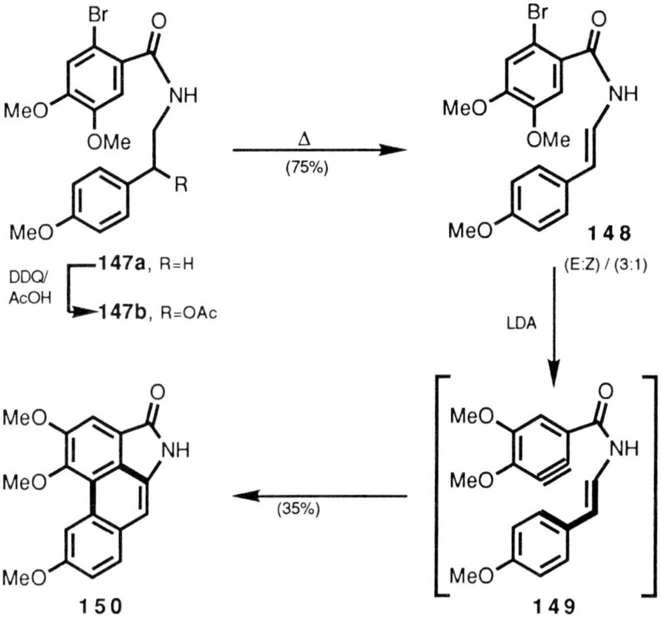

Scheme 37

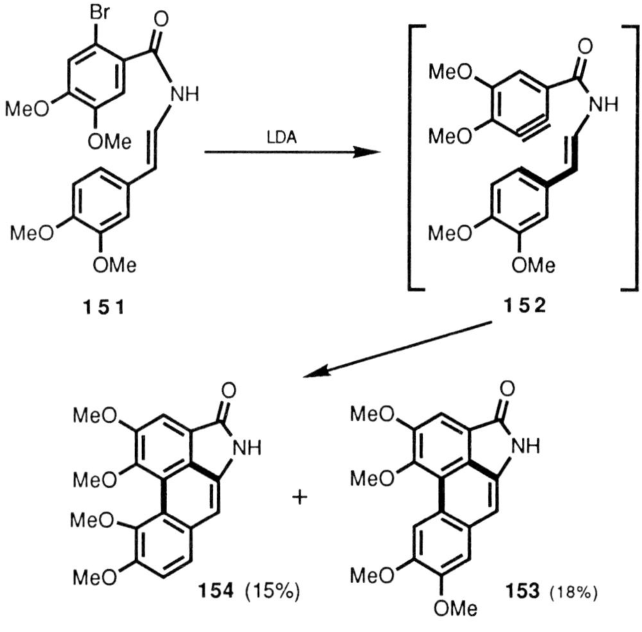

Scheme 38

We first investigated the synthesis of styrylamides **148**, which we were able to achieve simply and efficiently as follows. Oxidation of the readily prepared amide **147a** with DDQ in acetic acid gave a quantitative yield of the corresponding O-acetylderivative **147b**. Pyrolitic elimination at 250°C afforded the desired styrylamide **148** in 75% yield as a 3:1 mixture of E and Z isomers. Treatment of the former with LDA at 0°C afforded via intermediate **149**, a 35% yield of aristolactam **150**. When the E isomer of the unsymetrically substituted styrylamide **151** (Scheme 38), prepared as above, was treated with LDA under the same conditions, a mixture of the expected aristolactams **153** and **154** in 15% and 18% yields respectively, was obtained.[27]

Having achieved these results, we took on the challenge of further extending this approach to the synthesis of Amarillydaceae and indole-type alkaloids. Retrosynthetic analysis of the basic licorine skeleton **(156)** (Scheme 39), which is present in alkaloids such as anhydrolicorinium chloride **(155)**, led us to examine the possibility of synthesizing this type of Amarillydaceae alkaloid, which is known to possess antitumor activity, by a route in which the key step is the simultaneous formation of strategic bonds "a" and "b" by an intramolecular Diels-Alder reaction between the aryne and azadiene components contained in the intermediate **157**.

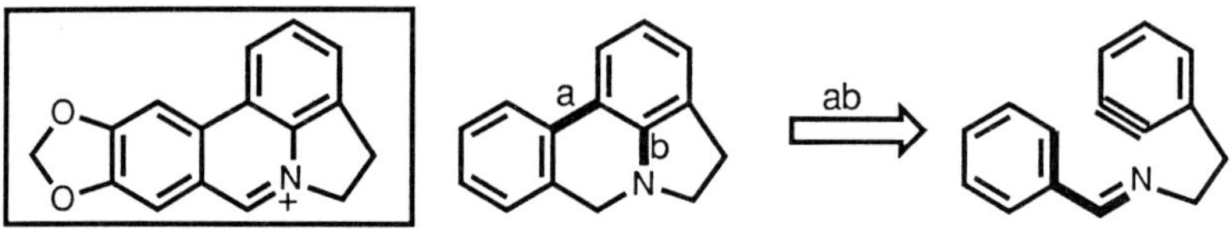

155   **156**   **157**

Scheme 39

To test this possibility we first prepared the amide **158** by simple condensation of 3,4-dimethoxybenzoic acid with 2-bromo-3,4-dimethoxy phenethylamine. When the amide **158** was treated with a large excess (16 eq.) of LDA, a 74% yield of the tetracyclic amide **162** was obtained (90% excluding recovered starting material); the key intermediate was the amide anion **161**, resulting from an intramolecular Diels-Alder reaction between the aryne and the diene system present in the amide anion **159** (Scheme 40). The regioisomer **163** was also obtained in this reaction, but in very small yield, presumably because the formation of cycloadduct **160** takes place via a more crowded transition state.[28]

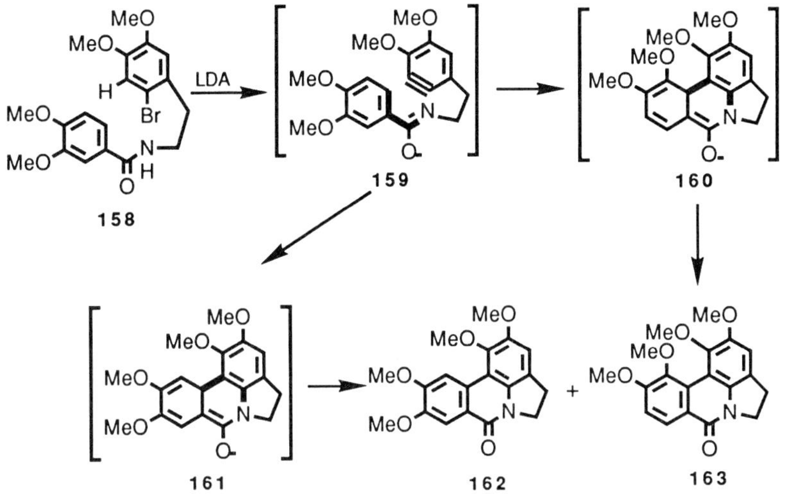

Scheme 40

Similar LDA treatment of imine **164**, which is easy to prepare by condensation of the corresponding aldehyde and amine, led via intermediates **165** and **166** to the same cyclic amide as above (**162**) (Scheme 41). Since compound **162** was not detected in the reaction mixture, it must have been produced by oxidation of the cycloadduct intermediate **166** during work-up.[29]

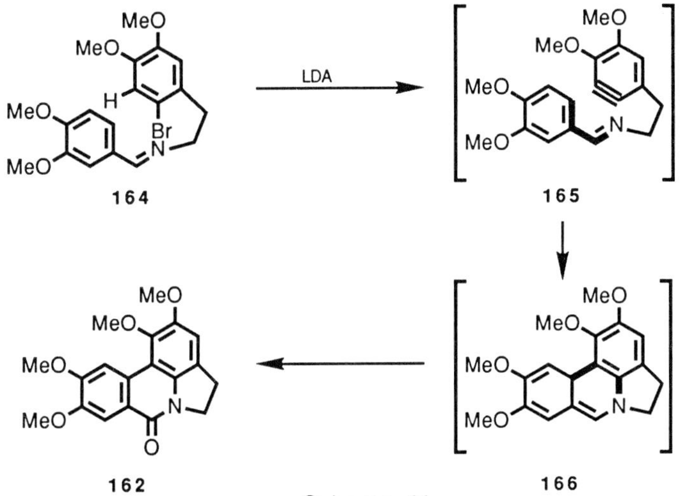

Scheme 41

As a preliminary to applying the above approach to the synthesis of indole alkaloids, we have studied the preparation of tetracyclic models such as **168**, which is closely related to the ergot alkaloids such as lysergic acid (**167**) (Scheme 42). Retrosynthetic analysis of compound **168** led us to investigate the possibility of simultaneously constructing bonds "a" and "b" by means of an intramolecular Diels-Alder reaction between aryne and styrene systems (**169**).

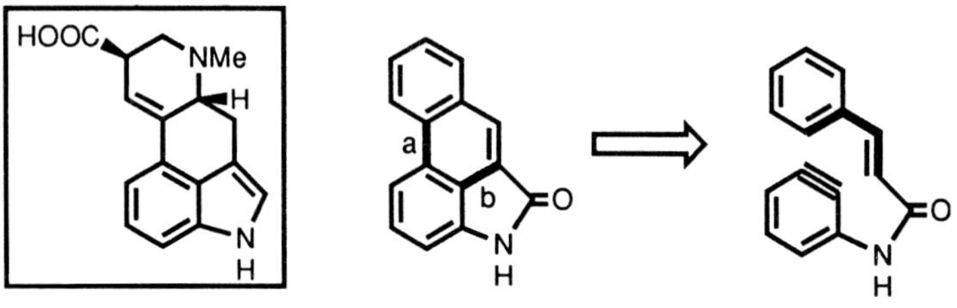

Scheme 42

Amide **170** was readily prepared by condensation of trans-cinnamoyl chloride and o-bromo-aniline. Preliminary studies have shown that treatment of amide **170** with a large excess of LDA in THF produced the expected compound **173** in 40% yield, formation of the intermediates **171** and **172** being followed by aromatization of the latter (Scheme 43).[30]

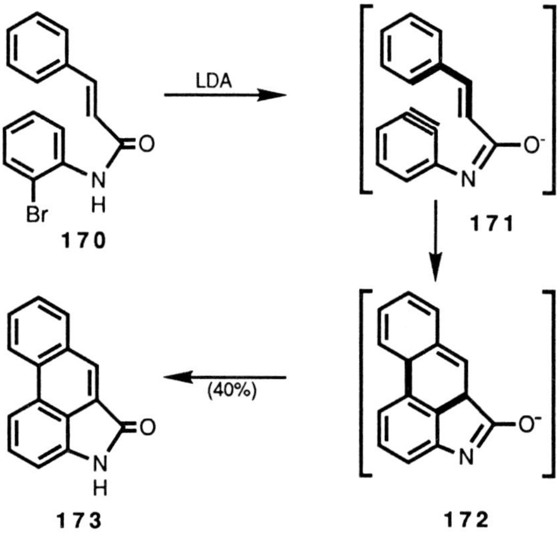

Scheme 43

4 INTRAMOLECULAR NUCLEOPHILIC ADDITION

Another interesting reaction of benzynes is nucleophilic addition. When this reaction is carried out intramolecularly it can provide cyclic compounds, that might be difficult to prepare by other means.

We have applied this methodology to the synthesis of the 9-substituted-9H-xanthenes **176**, which we are studying as synthetic precursors of the alkaloid clavizepine (**177**).[31] In our synthesis of the xanthenes **176**, aryne chemistry was used to form the pyran ring.[32] This was achieved by treatment of the bromoderivatives **174a**, **174b** and **174c** with LDA, which via the aryne intermediate **175** produced the corresponding cyclic compounds **176a**, **176b** and **176c** in 60%, 76% and 59% yields respectively.

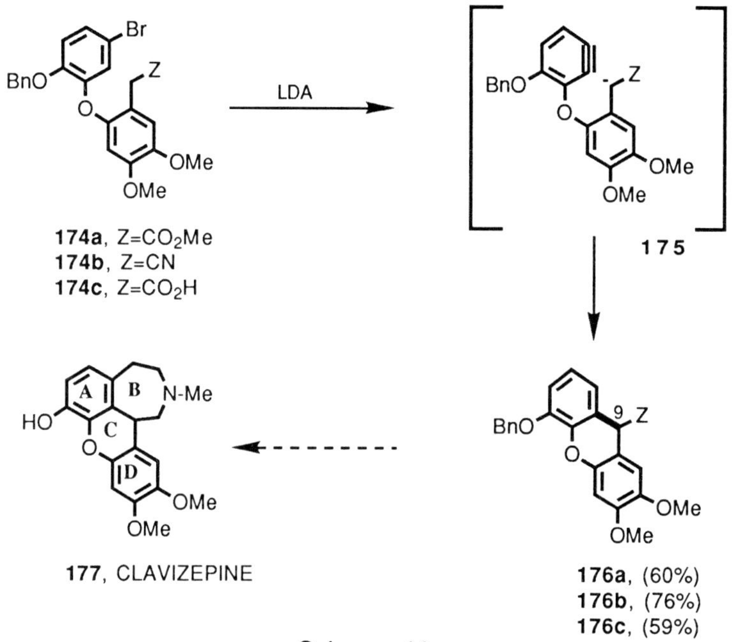

Scheme 44

Construction of ring B of clavizepine (**177**) is now being carried out by applying the usual methods to compounds **176**.

ACKNOWLEDGEMENT

We wish to thank our highly dedicated group of coworkers for their contributions to the research presented in this manuscript. We are also grateful to the CICYT and Consellería de Educación e Ordenación Universitaria (Xunta de Galicia) for their generous support of this research.

REFERENCES

1. M. Shamma, "The Isoquinoline Alkaloids, Chemistry and Pharmacology", Academic Press, 1972; M. Shamma and J.L. , Moniot, "Isoquinoline Alkaloid Research 1972-1977", Plenum Press, 1978; T. Kametani and T. Honda, in "The Alkaloids" (Arnold Brossi ed.), Vol. 24, p. 153, Academic Press, New York, 1985; Z-L. Chang and D-Y. Zhu, in "The Alkaloids" (Arnold Brossi ed.), Vol. 31, p. 29, Academic Press, New York, 1987, and L. Castedo and E. Guitián, in "Natural Products Chemistry III" (Atta-ur-Rahman and P.W. Le Quesne eds.), pp 235-245,Springer-Verlag, New York, 1988.
2. L. Castedo, E. Guitián, J.M. Saá and R. Suau, Tetrahedron Letters, 1982, 23, 457.
3. A. Brossi, J. Würsch and O.Schneider, Chimia, 1958, 12, 114.
4. A. Brossi, L.A. Dolan and S. Teitel, Org.Synth., 1977, 56, 3.
5. G.R. Lenz, C.M. Woo and B.L. Hawkins, J.Org.Chem., 1982, 47, 3049.
6. C. Saá, PhD. Thesis, Universidad de Santiago de Compostela (Spain), 1985.
7. N. Atanes, PhD. Thesis, Universidad de Santiago de Compostela (Spain), 1988.
8. L. Friedman and F.M. Logullo, Org.Synth., 1969, 48, 12.
9. C. Saá, E. Guitián, L.Castedo and J.M. Saá, Tetrahedron Letters, 1985, 26, 4559.
10. B. Gómez, G. Martín, E. Guitián, L. Castedo and J.M. Saá, unpublished results.
11. E. Guitián, L. Castedo, J.M. Saá and R. Suau, unpublished results.
12. L. Castedo, J.R. Estévez, J.M. Saá and R. Suau, Tetrahedron Letters,1978, 19, 2179, and L. Castedo, J.R. Estévez, J.M. Saá and R. Suau, J.Heterocycl.Chem., 1982, 19, 1319.
13. L. Castedo, E. Guitián, C. Saá, R. Suau and J.M. Saá, Tetrahedron Letters, 1983, 24, 2107.
14. N. Atanes, L. Castedo, E. Guitián and J.M. Saá, Heterocycles, 1987, 26, 1183.
15. N. Atanes, E. Guitián, C. Saá, L. Castedo and J.M. Saá, Tetrahedron Letters, 1987, 28, 817
16. N. Atanes, L. Castedo, A. Cobas, E. Guitián, C. Saá and J.M. Saá, Tetrahedron, 1989, 45, 7947.
17. C. Saá, E.Guitián, L.Castedo, R. Suau and J.M. Saá, J.Org.Chem., 1986, 51, 2781.
18. A. Cobas, E. Guitián, L. Castedo and J.M. Saá, Tetrahedron Letters, 1988, 29, 2491.
19. L. Castedo, A. Cobas, E. Guitián and J.M. Saá, "Sixth FECHEM Conference on Heterocycles in Bio-Organic Chemistry", Abstracts of posters, Solbacka, Sweden, 1990, 5.
20. L. Castedo, A. Cobas, E. Guitián and J.M. Saá, unpublished results.
21. N. Atanes, E. Guitián, L. Castedo and J.M. Saá, unpublished results.
22. G. Martín, E. Guitián, L. Castedo and J.M. Saá, Tetrahedron Letters, 1987, 28, 2407.
23. G. Martín, E. Guitián, L. Castedo and J.M. Saá, unpublished results.
24. S.F. Dyke, M. Sainsbury, D.W. Brown and R. Clipperton, Tetrahedron, 1970, 26, 5969.

25. D. Pérez Meirás, E. Guitián and L. Castedo, Tetrahedron Letters, 1990, 31, 143.
26. D. Pérez Meirás, E. Guitián and L. Castedo, unpublished result.
27. J.C. Estévez, R.J. Estévez, E. Guitián, M.C. Villaverde and L. Castedo, Tetrahedron Letters, 1989, 30, 5785.
28. D. Pérez Meirás, E. Guitián and L. Castedo, Tetrahedron Letters, 1990, 31, 2331.
29. C. González Bello, E. Guitián and L. Castedo, unpublished results.
30. B. Gómez Pérez, E. Guitián and L. Castedo, unpublished results.
31. J. Boente, L. Castedo, D. Domínguez and M. Ferro, Tetrahedron Letters, 1986, 27, 4077.
32. R. Vázquez, L. Castedo, and D. Domínguez, "Sixth FECHEM Conference on Heterocycles in Bio-Organic Chemistry", Abstracts of posters, Solbacka, Sweden, 1990, 7.

New Synthetic Approaches to Nuphar Alkaloids

J. Nowacki and J. T. Wróbel

DEPARTMENT OF CHEMISTRY, THE UNIVERSITY OF WARSAW, 02-093 WARSAW, POLAND

Nuphar alkaloids have been isolated from species belonging to the genus Nuphar of the Nymphaeaceae family (widely known as water lilies). A few structurally related compounds have been obtained from the scent glands of the Canadian beaver.

The characteristic structural feature, common for all these natural products, is the quinolizidine - sesquiterpenoid skeleton substituted in positions 1,4 and 7. The conformation of the quinolizidine ring system is generally trans except for nupharidine where it is *cis* and for nupharopumiline where the sterochemistry is dependent on the presence of the double bond. The relative configuration of the characteristic 3-furyl group at carbon 4 is always equatorial.

Nuphar alkaloids have for the last forty years been the subject of detailed investigations and the results have been systematically reviewed[1]. In spite of the fact that major progress in their synthesis has been made,[2-12] several functionally complex molecules, like the sulfurcontaining Nuphar alkaloids, have not yet been synthesized.

A detailed survey of the synthetic procedures reported in the lite-rature reveals that the efforts were aimed exclusively at the preparation of selected individual Nuphar alkaloids. This restricted strategy determined the choice of the starting materials, the structure of the intermediates as well as the synthetic pathways.

In order to develop a general approach to Nuphar alkaloids we have synthesized a number of 4-quinolizidone derivatives with various substituents in position 1 and 7. The goal was achieved using the reaction sequence given in *Scheme 1* starting with commercially available 2-piperidone or simple derivatives (**1**, R=CH$_3$; CO$_2$Et; OAc).

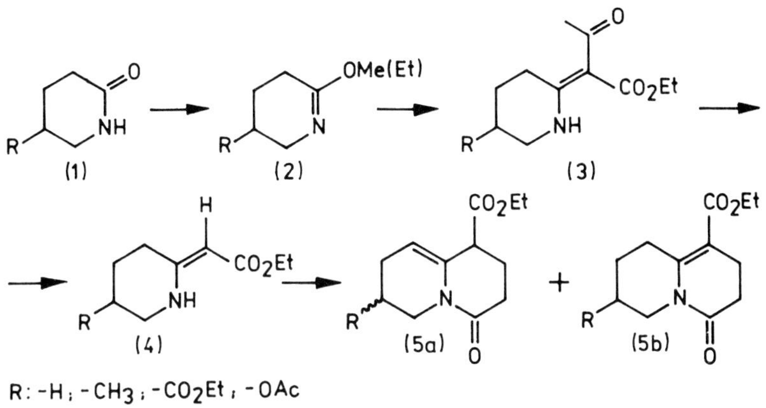

Scheme 1

The products **5a** and **5b** (always obtained as tautomeric mixtures#) were separated by liquid chromatography.

Evaluation of the synthetic potential of these new cyclic enamidoesters (*Scheme 2*) afforded some very promising advanced precursors of Nuphar alkaloids such as **14, 15, 18, 20**, as well as the julolidine derivatives **24, 25, 27**, and a number of novel intriguing quinolizidine derivatives, e.g. **12, 13, 22, 23** and **28**. Conversion of the two (vide infra) primary reduction products of (**5b**; R=H) with sodium borohydride (NBH), i.e. **8** and **10**, resulted in (+)-lupinin **9** and an azaanalogue of the olive fly pheromone **11**[13]. The dieneamides **18** were found to be very sensitive substrates in photooxidation reactions.

The results obtained encouraged us to explore intensively the scope of these reactions. Attempts to rationalize the synthetic data lead to the conclusion that a feature in the reaction mechanisms of cyclic enamidoesters is the intermediacy of N-acyliminium ions as shown in *Scheme 3*.

#Only one isomer is formed in an analogous reaction sequence starting from 2-pyrrolidone.

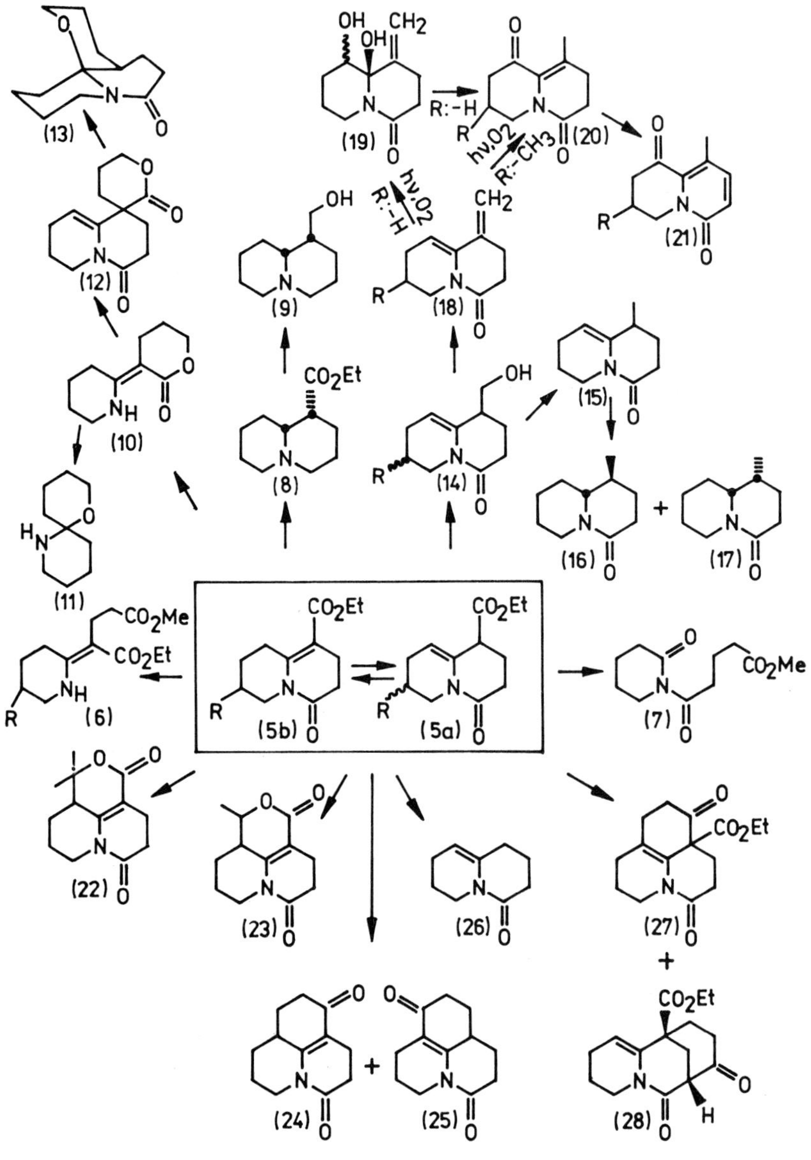

Scheme 2

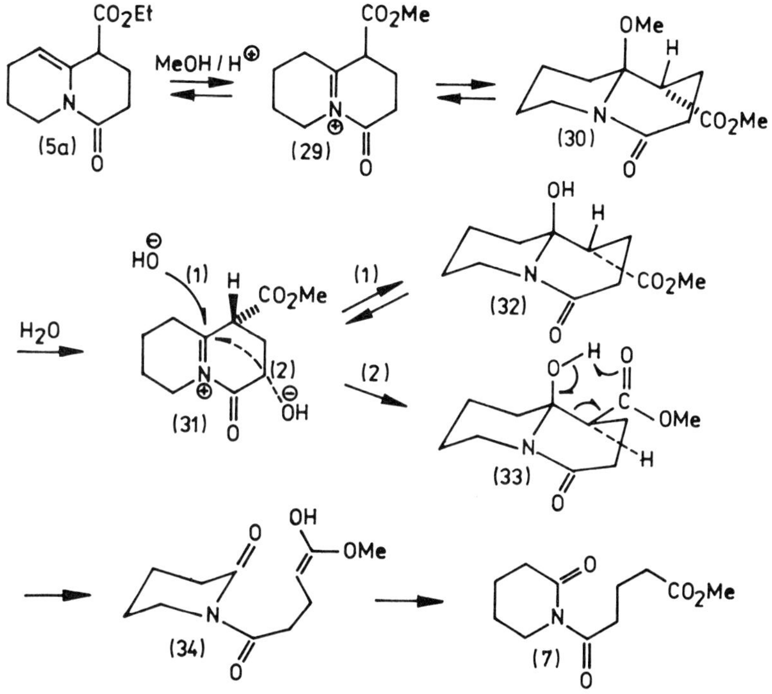

Scheme 3

The outcome of the reduction (NBH) of cyclic enamidoesters depended on the type of alcohol used *(Scheme 4)*. Analysis of this phenomenon induced us to postulate that the crucial factor is the solvent-intermediate 36 interaction *(Scheme 5)*. This in turn depended on the type of alcohol (primary, secondary or tertiary) used.[14]

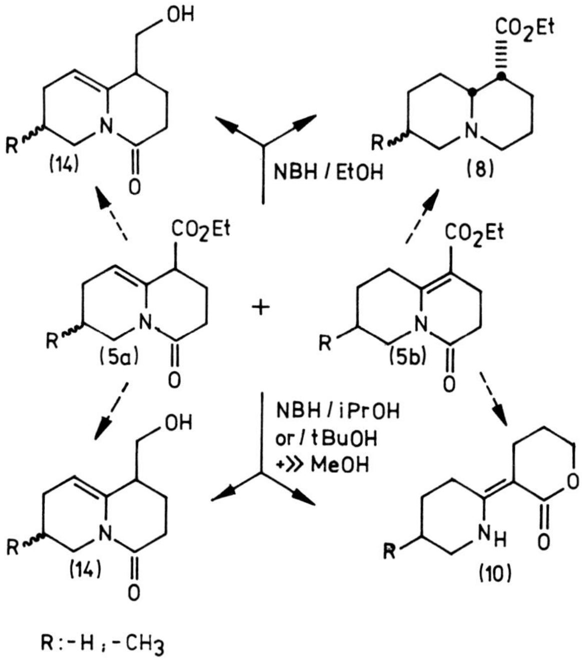

Scheme 4

A parallel study is focused on testing model synthetic routes in order to elaborate procedures effective for conversion of available functional groups on carbon 7 of cyclic enamidoesters into functions and structural arrangements occurring in Nuphar alkaloids. The most challenging remaining problem is the construction of the spiro junction between the quinolizidine and tetrahydrothiophene rings. Scheme 6 is intended to give an impression of how we will tackle this problem applying 7-oxoquinolizidine-4-one **40** as a model.

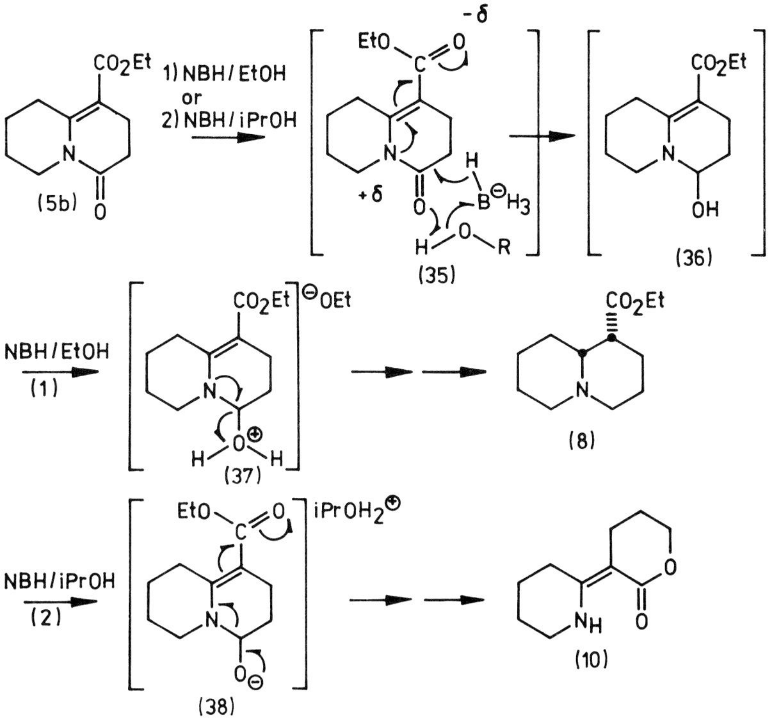

Scheme 5

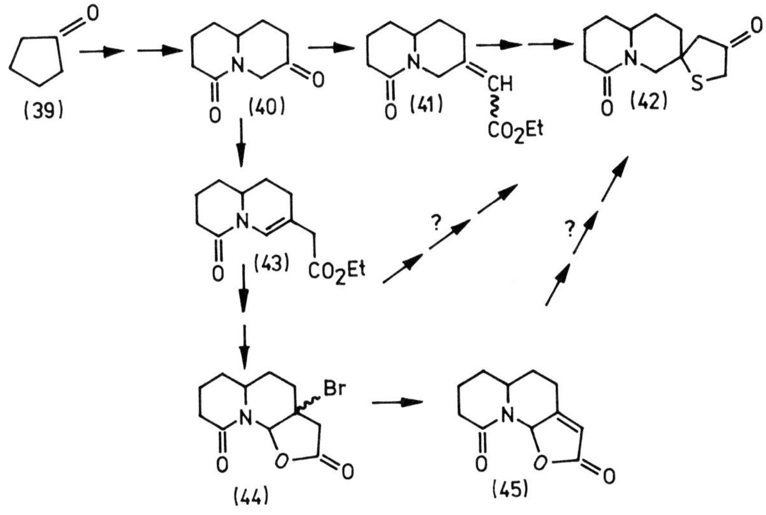

Scheme 6

REFERENCES

1. J.T. Wróbel, "The Alkaloids" R.H.F. Manske Academic Press, New York, 1967, Vol. 9, Chapter 10, p. 441.;
   J.T. Wróbel, ibid. 1977, Vol. 16, Chapter 3, p. 181;
   J.T. Wróbel, ibid. 1989, Vol 35, Chapter 5, p. 215.
2. M. Kotake, I. Kawasaki, T. Okamoto, S. Kusumoto, and T. Kaneko, Ann. Chem., (1960), **636** 158;
   T. Kaneko, I. Kawasaki and T. Okamoto, Chem. Ind., (1959), 1191.
3. M. Kotake, I. Kawasaki, S. Matsutani, S. Kusumoto and T. Kaneko, Bull. Chem. Soc. Japan, (1962), **35**, 1494.
4. Y. Arata and T. Nakanishi, J. Pharm. Soc. Japan, (1960), **80**, 855;
   Y. Arata, T. Nakanishi and Y. Asaoka. Chem. Pharm. Bull. (1962), **10**, 675.
5. I. Jezo, M. Karwas and K. Tihlarik, Chem. Zvesti, (1961), **15**, 283.
6. J.T. Wróbel and Z. Dabrowski, Roczniki Chemii, (1965), **39**, 1239.
7. F. Bohlmann, E. Winterfeldt, H. Laurent and W. Ude, Tetrahedron, (1963), **19**, 195.
8. J. Szychowski, A. Leniewski and J.T. Wróbel, Chem. Ind., (1978), 273.
9. R.T. LaLonde, N. Muhammad, C.F. Wong and E.R. Sturiale,. J. Org. Chem., (1980), **45**, 3664.
10. S. Yasuda, M. Hanaoka and Y. Arata, Chem. Pharm. Bull, (1980),**28**, 831.
11. M. Natsume and M. Ogawa, Heterocycles, (1981), **15**, 237.
12. Y.C. Hwang and F.W. Foowler, J. Org. Chem., (1985), **50**, 2719.
13. R. Baker, R. Herbert, P.E. Howse and O.T. Jones, J.C.S. Chem. Comm., (1980), 52.
14. M.J. Kamlet, J.L.M. Abboud and R.W. Taft, Progr. Phys. Org. Chem., (1981), **13**, 485..

Antiallergic 4H-Pyrido[1,2-a]pyrimidin-4-ones: Synthesis and Structure Activity Relationship

I. Hermecz

RESEARCH CENTRE, CHINOIN PHARMACEUTICAL AND CHEMICAL WORKS LTD., BUDAPEST, TÓ U. 1-5. H-1045 HUNGARY

In the last two decades among 4H-pyrido[1,2-a]pyrimidin-4-ones numerous derivatives exhibited outstanding pharmacological activities[1]. This presentation gives account of the research on the antiallergic 4H-pyrido[1,2-a]pyrimidin-4-ones carried out in Chinoin. Initially, our research was directed towards the synthesis of pyrido[1,2-a]pyrimidinones having CNS effects, primarily an analgetic effect[2,3], but also antiinflammatory[4,5] and antithrombotic activities[6] (Scheme 1).

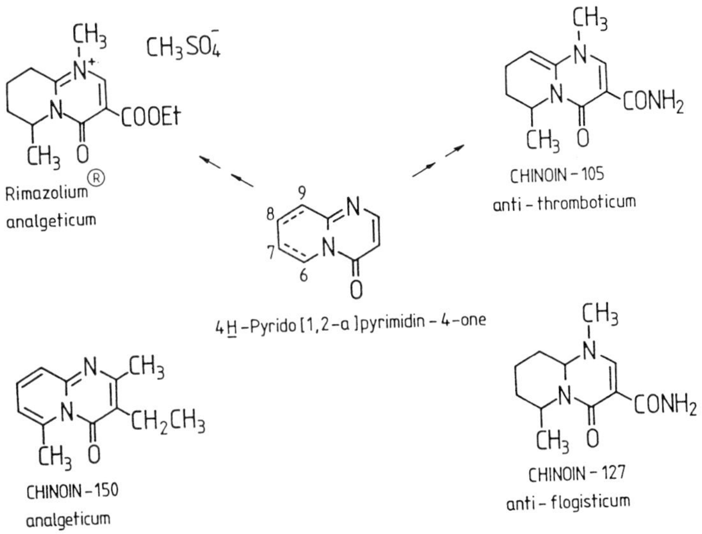

Scheme 1.

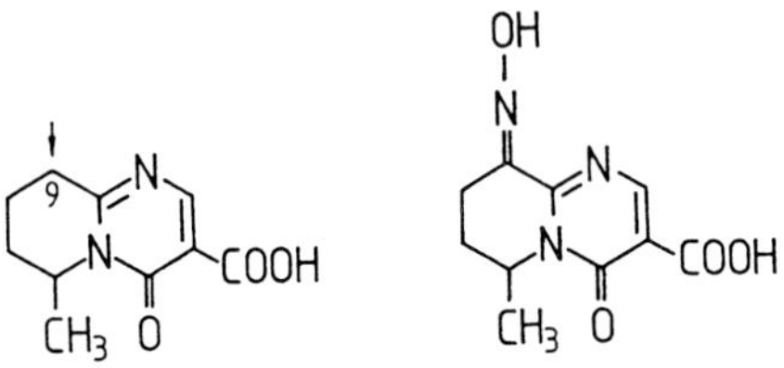

Scheme 2.

PCA test
ID_{50} 240 μMol/kg iv. ID_{50} 19.4 μMol/kg

During structure-activity relationship studies, it was demonstrated that the starting 6,7,8,9-tetrahydropyrido-pyrimidinones contain[7] a reactive methylene group at position 9, and introduction of a substituent into this position resulted in loss of the CNS effects[8]. At the same time, favourable new pharmacological effects appeared. For example, the carboxylic acid exhibited a very weak antiallergic activity in rats in the passive cutaneous ana-phylaxis test (PCA). If an isonitroso group is present at position 9, the compound has no CNS effects, but exhibits a moderate PCA activity[9] (Scheme 2). It is roughly ten times more potent than the parent carboxylic acid. After this observation, we elaborated a project to study systematically the reactions of the active methylene group, hoping that the chemical transformations in this position would lead to more active antiallergic derivatives (Scheme 3).

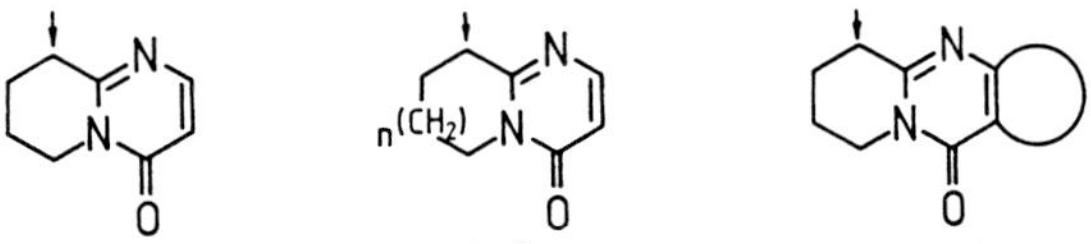

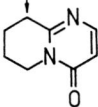

n = 0–3

Reactions of
Active Methylene Group

Halogenation (Bromination)

Diazonium coupling
Vielsmeyer-Haack formylation

Products were reacted with

N-nucleophiles: Hydrazines
Amines
Azide
Nitrites

Fischer-indolization
Amines
Japp-Klingeman reaction
Diazomethane
Clayfen

Scheme 3

The lecture deals with the bromination, diazonium coupling and Vielsmeier-Haack formylation of tetrahydropyridopyrimidinones. Bromo derivatives were reacted with N-nucleophiles, such as hydrazines, amines, azide and nitrite. The lecture also deals with the Fischer indolization of hydrazono derivatives. The formyl derivatives obtained in the Vielsmeier-Haack reaction were reacted with amines, diazonium salts (Japp-Klingeman reaction), diazomethane and clay-fen (a clay-supported ferric-nitrate).

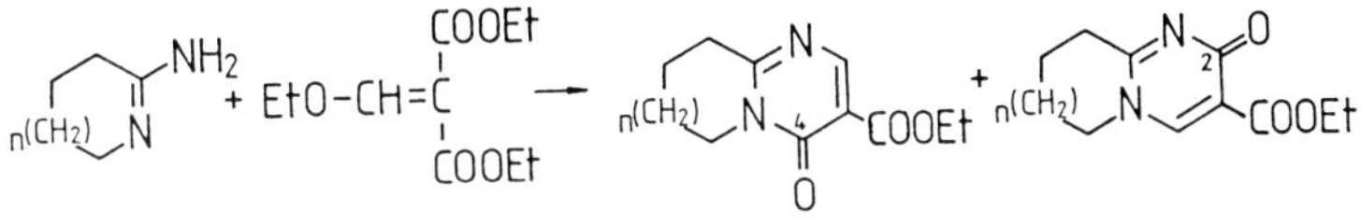

n		
n = 0	69 %	16 %
n = 1	72 %	25 %
n = 2	80.5%	8.9 %
n = 3	80 %	4.8 %
n = 4	72 %	?

Scheme 4.

Tetrahydropyridopyrimidinones and their piperidine ring homologues can readily be prepared, for example in the reaction of semicyclic amidines with diethyl ethoxymethylenemalonate[10] (Scheme 4). The ring-closure reactions proceeded smoothly at room temperature in ethanol. The 4-oxo derivatives were accompanied by 2-oxo isomers. The yields of the latter decreased with increasing ring size. For the azonine derivative, the presence of the 2-oxo isomer could be only detected by means of thin-layer chromatography, but we failed to isolate it.

We estimated the reactivity of the active methylene groups of the nitrogen bridgehead compounds via hydrogen-deuterium exchange (Scheme 5). The hydrochloride of the investigated compound was dissolved in deuterium oxide and the decrease in intensity of the protons of the active methylene group was followed by proton NMR spectroscopy. The hydrogen-deuterium exchange probably involves the imine-enamine tautomerism. If the kinetic acidity of the active methylene group of the five-membered derivative is taken as unit, that in the six-membered homologue is about 64 times more active, while those in the higher homologues are only one-tenth as active. These differences in hydrogen-deuterium exchange probably stem from the conformation differences in the saturated ring of the bicyclic compounds. Among the 4-oxo-4H-pyrido[1,2-a]pyrimidine-3-carboxylates, the 6-methyl derivative contains the most reactive methylene group. The reactivity of the methylene group is also

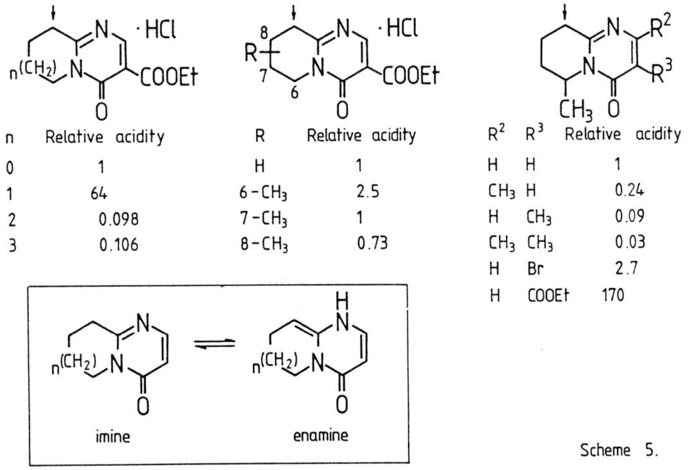

n	Relative acidity
0	1
1	64
2	0.098
3	0.106

R	Relative acidity
H	1
6-CH_3	2.5
7-CH_3	1
8-CH_3	0.73

R^2	R^3	Relative acidity
H	H	1
CH_3	H	0.24
H	CH_3	0.09
CH_3	CH_3	0.03
H	Br	2.7
H	COOEt	170

imine ⇌ enamine

Scheme 5.

affected by the nature of the substituents at positions 2 and/or 3. An electron-donating group, for example a methyl group, decreases the activity of the active methylene group, while an electron-withdrawing group increases it. The active methylene group is more sensitive to the substituent present at position 3 than to that at position 2. Thus an electron-donating methyl group at position 2 decreases the acidity to one-fourth compared with that of the parent bicycle, but to one-tenth if it is present at position 3.

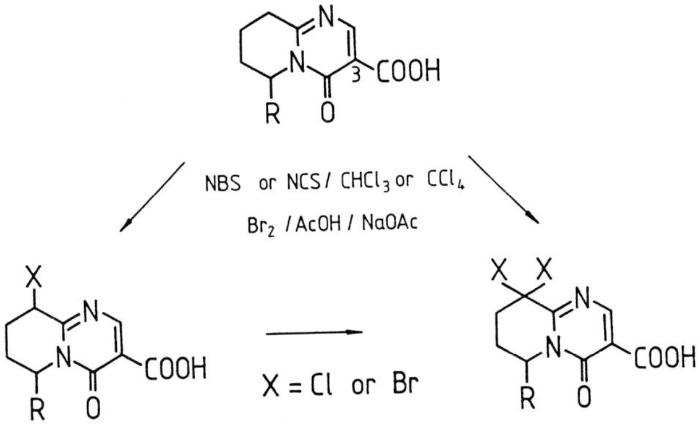

NBS or NCS / $CHCl_3$ or CCl_4
Br_2 / AcOH / NaOAc

X = Cl or Br

Scheme 6.

Scheme 6 demonstrates the halogenations of the pyridopyrimidine-3-carboxylic acids with N-bromosuccinimide or N-chlorosuccinimide in chloroform or carbon tetrachloride, or with bromine in acetic acid in the presence of sodium acetate[3,11,12]. Depending on the molar ratio of the substrate and reactant, mono- or dihalogenated derivatives could be obtained. Under similar reaction conditions, the monohalogenated derivatives could be converted into the disubstituted compounds.

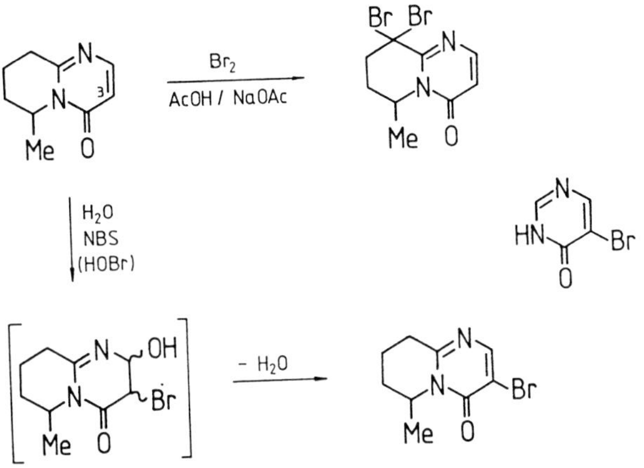

Scheme 7.

In the case of 3-unsubstituted tetrahydropyridopyrimidines regioselective halogenation was observed (Scheme 7). For example, when the 6-methyl derivative was reacted with bromine in acetic acid, the dibromo compound was obtained, but if the bromination was carried out in water with N-bromosuccinimide, the 3-bromopyridopyrimidine could be isolated. In the latter case, processes of addition and elimination of hypobromous acid, formed from NBS on the action of water, probably took place at the C(2)-C(3) double bond. A similar mechanism was suggested by Tee and Banerjee for the bromination of monocyclic 3H-pyrimidin-4-one by NBS in water[13].

The reactions of 9,9-dibromopyridopyrimidines with hydrazine afforded 9-hydrazono derivatives in good yields[14]. As the 9-hydrazonopyridopyrimidine-3-carboxylic acids possessed high antiallergic activities, further syntheses of 9-hydrazono derivatives were elaborated, especially from a patenting point of view (Scheme 8). The active methylene group readily reacted with diazonium salts and the appropriate hydrazones were obtained[15,19].

When the 9-bromopyridopyrimidines were reacted with hydrazines, again the hydrazono derivatives were the products[16]. That is, besides substitution, an oxidation process took place. The hydrazono derivatives were likewise the products when the reactions were

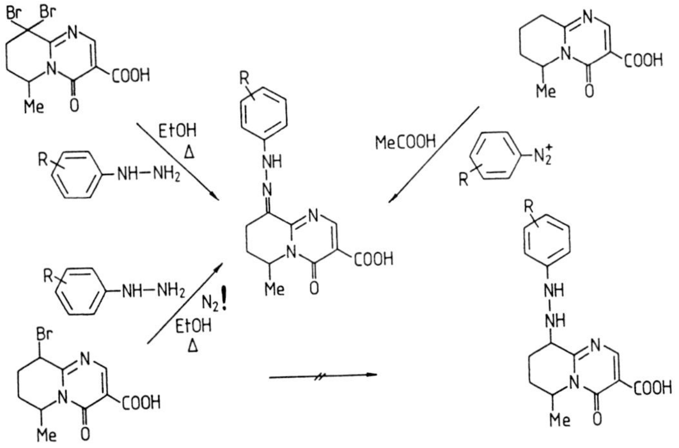

Scheme 8.

carried out under an inert atmosphere. The best yields could be achieved if two mols of hydrazine were applied per mol of 9-bromo derivative. A careful analysis of the reaction mixture indicated that one mol of the appropriate aniline and one mol of ammonium bromide were also formed. On the basis of these data, we suggest the following osazone-like mechanism, which is well known in sugar chemistry (Scheme 9). In the first step substitution occurs. Then, in the tautomeric 1,6,7,8-tetrahydro form, cleavage of the N-N bond takes place to give an amine and 9-iminopyridopyrimidine. In the next step, the latter product reacts with a second mol of hydrazine to afford the 9-hydrazonopyridopyrimidine and ammonia.

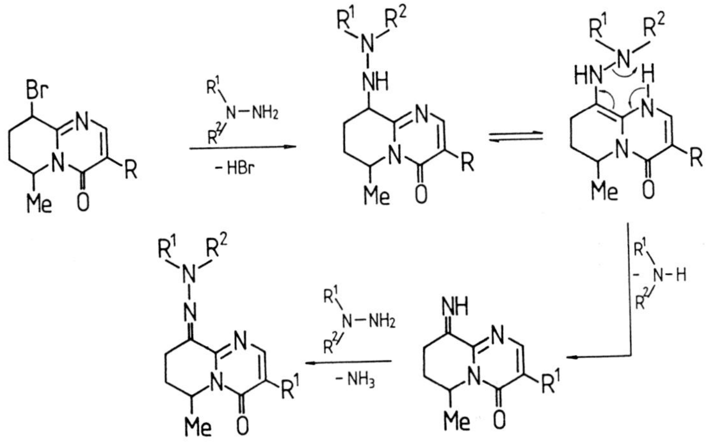

Scheme 9.

For the structures of the 9-hydrazonopyridopyrimidines, different tautomeric forms may be considered, but only the presence of the hydrazono form could be detected when different spectroscopic methods were used[14,18,21] (Scheme 10).

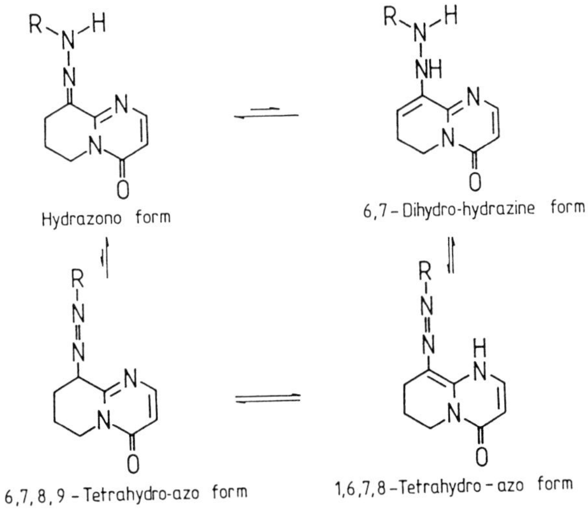

Hydrazono form

6,7-Dihydro-hydrazine form

6,7,8,9-Tetrahydro-azo form

1,6,7,8-Tetrahydro-azo form

Scheme 10.

Due to the presence of the C = N double bond, the hydrazones may exist as E and/or Z geometric isomers. In fact, we observed a solvent-dependent geometric isomerism, which points to the low activation energy of interconversion between the two geometric forms (Scheme 11). In deuterochloroform, the sterically crowded Z isomers predominate because of a hydrogen-bond between the nitrogen atom at position 1 and the amino group. In hexadeuterodimethyl sulfoxide, where a stronger hydrogen-bond could form with the solvent, the sterically more favourable E geometric isomers predominate[14,19]. However, when a methyl group was present at position 8, the steric properties were balanced between two geometric forms and again the Z isomer was the predominant. Because of no possibility for the formation of hydrogen bond, the N-methyl derivative exists as the E geometric isomer independently of the nature of the solvent.

Fischer indolization of the arylhydrazono derivatives could readily be carried out in polyphosphoric acid at 180 ^{0}C, to give tetracyclic indolo derivatives (Scheme 12).

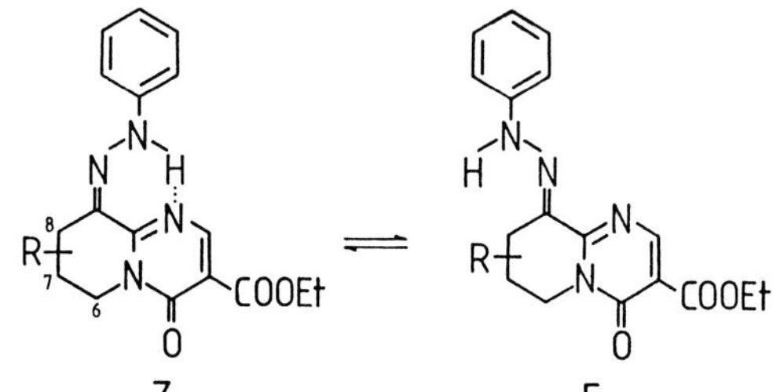

R	in CDCl$_3$		in DMSO-d$_6$	
	E %	Z %	E %	Z %
H	6	94	75	25
6-CH$_3$	20	80	83	17
7-CH$_3$	5	95	83	17
8-CH$_3$	0	100	25	75

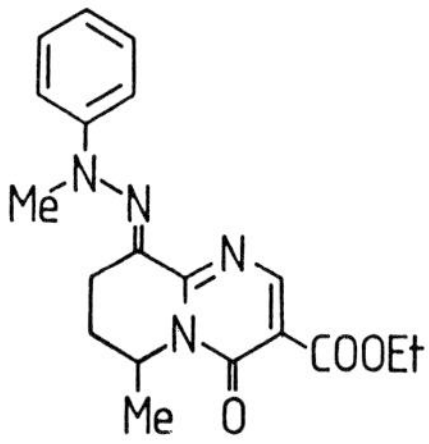

Scheme 11.

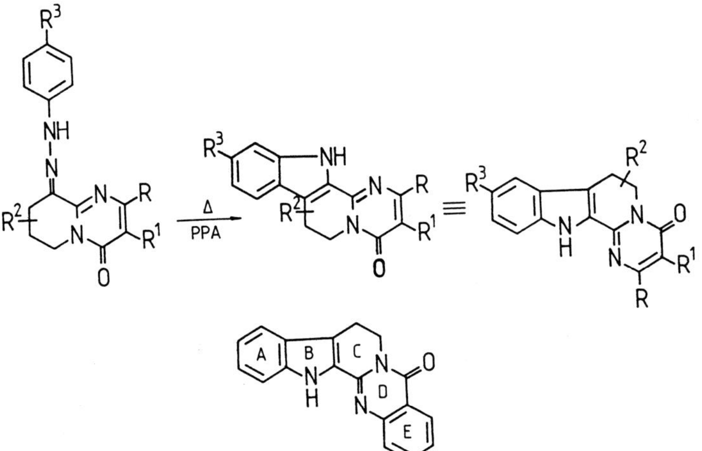

Rutaecarpine

Scheme 12.

This reaction could be applied for the total synthesis of rutaecarpine alkaloids[17,18,20]. The common principle with the earlier rutaecarpine synthesis is to build up the connection of rings C and D, starting from tryptamine or its derivatives (Scheme 13).

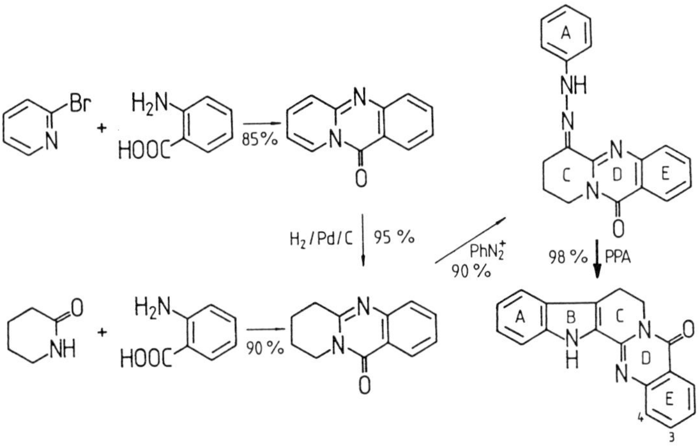

Scheme 13.

In the last step of our method, the connection of rings A and C is formed. This method could conveniently be extended to the peparation of rutaecarpine derivatives substituted in rings A, C or E, and also of 1,2,3,4-tetrahydro derivatives and C or E ring homologues, using the appropriate starting materials.

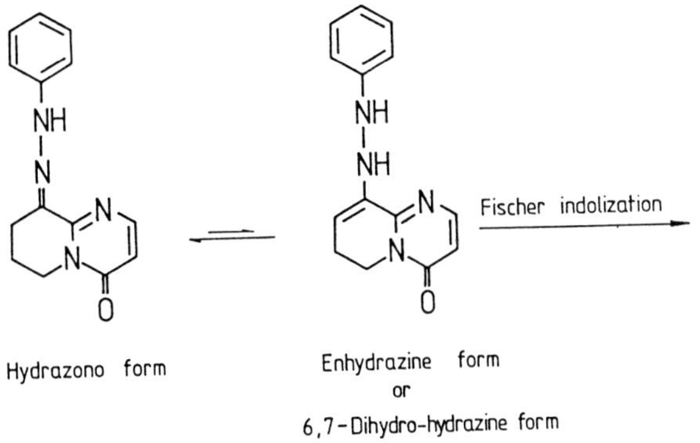

Hydrazono form

Enhydrazine form
or
6,7-Dihydro-hydrazine form

Scheme 14.

The mechanism of the Fischer indolization of arylhydrazones was investigated in great detail by other authors, and it was unambiguously confirmed that the reaction took place from the enhydrazine tautomeric form[24,25] (Scheme 14).

We earlier saw that in our case only the hydrazono form could be detected. We wondered whether the equilibrium between the 6,7,8,9-tetrahydro and 6,7-dihydro forms could be shifted to the latter form by modification of the structure to such an extent that the 6,7-dihydro form could also be detected.

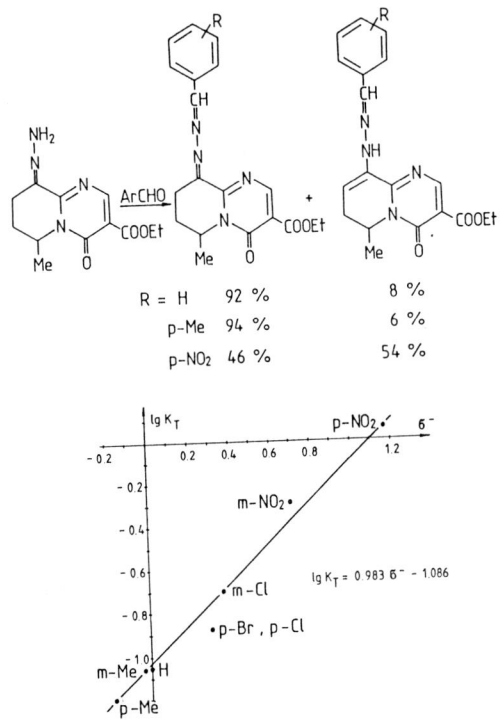

Scheme 15.

We met success when 9-hydrazonopyridopyrimidine was reacted with benzaldehyde (Scheme 15). In the case of benzaldehyde the product was present in equilibrium as a 9:1 mixture of the tetrahydro and dihydro tautomers in deuterochloroform. The ratio of the two tautomeric forms was affected by the substitutent on the phenyl ring. We obtained a fair linear regression between the equilibrium constant and the 6^- constant of the substituent present on the phenyl ring.

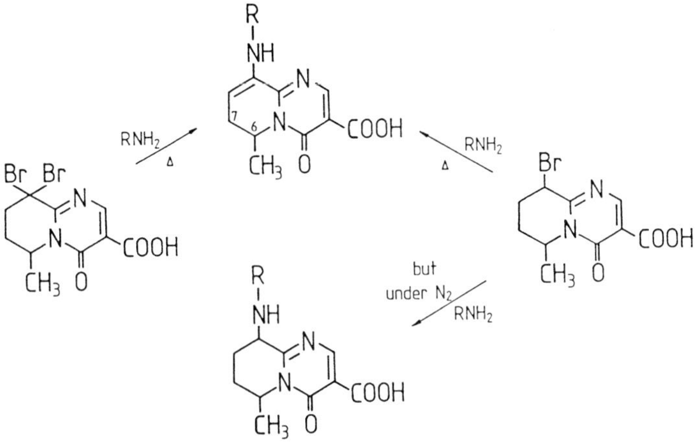

Scheme 16.

We reacted the 9-bromo- and dibromopyridopyrimidines with amines, too[9,26,27]. In both cases, 9-amino-6,7-dihydropyridopyrimidines were the products (Scheme 16). That is, the substitution was again accompanied by an oxidation procedure when we started from 9-bromo derivatives. However, if the reaction was carried out under an inert atmosphere, in contrast to the reaction with hydrazine, 9-aminotetrahydropyridopyrimidines could be isolated[28].

Characteristic UV data of
9 - aminotetrahydropyrido [1,2-a]pyrimidinones

1st group	2 nd group
238 - 247 ($\varepsilon \sim 14\,000$)	265 ($\varepsilon \sim 30\,000$)
280 - 300 ($\varepsilon \sim 7\,000$)	355 - 365 ($\varepsilon \sim 5\,000$)
220 - 230 ($\varepsilon \sim 7\,000$)	255 - 260 ($\varepsilon \sim 30\,000$)
280 - 305 ($\varepsilon \sim 8\,000$)	340 - 365 ($\varepsilon \sim 3\,000$)

Scheme 17.

The ethanolic UV spectra of the 9-aminotetrahydropyridopyrimidinones could be divided into two groups[29] (Scheme 17). In the first group, the UV spectra of the 9-amino derivatives were very similar to those of 6,7,8,9-tetrahydropyridopyrimidinones[10], while in the second group the UV spectra of the products resembled those of 1,6,7,8-tetrahydropyrimidinones[27].

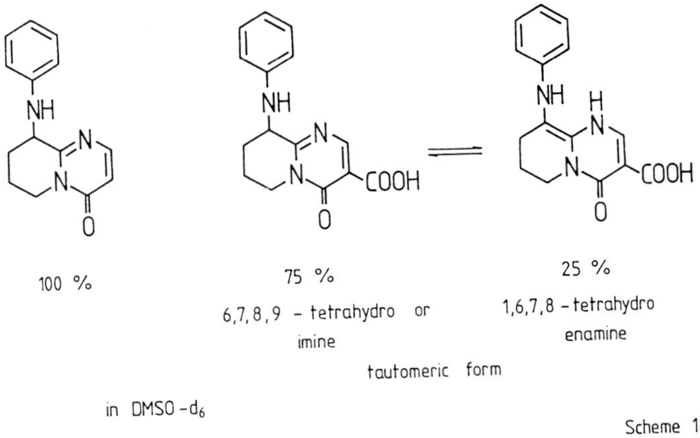

100 %

75 %
6,7,8,9 - tetrahydro or imine

25 %
1,6,7,8 - tetrahydro enamine

tautomeric form

in DMSO-d_6

Scheme 18.

On the basis of the UV spectra, we supposed that the 9-aminotetrahydropyridopyrimidinones might exist as 6,7,8,9- and 1,6,7,8-tetrahydro tautomers. This supposition was justified by detailed ^{1}H and ^{13}C NMR investigations[28].

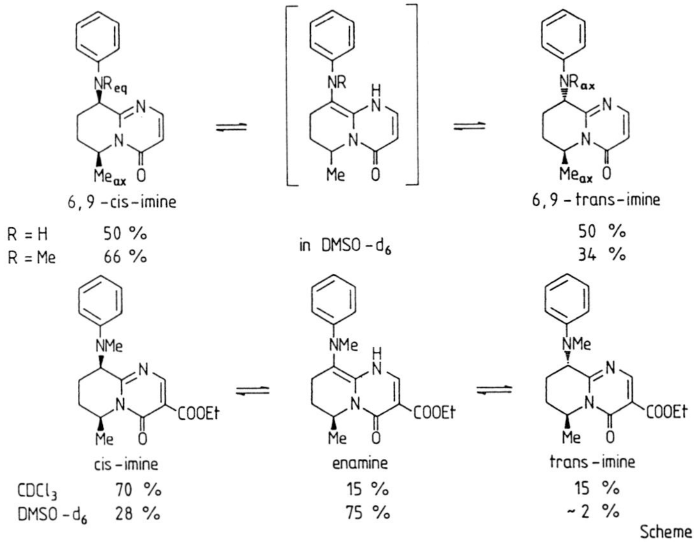

6,9-cis-imine
R = H   50 %
R = Me   66 %

in DMSO-d_6

6,9-trans-imine
50 %
34 %

cis-imine
CDCl$_3$ 70 %
DMSO-d_6 28 %

enamine
15 %
75 %

trans-imine
15 %
~ 2 %

Scheme 19.

9-Phenylaminotetrahydropyridopyrimidinones exist only in the 6,7,8,9-tetrahydro tautomeric form in hexadeuterodimethyl sulfoxide, while the 3-carboxylic acid is present as a 3:1 mixture of imine and enamine tautomers (Scheme 18).

If a methyl group is present at position 6, cis and trans isomers can be identified within the imine tautomer (Scheme 19). In both isomers, the methyl group at position 6 always occupies the quasi-axial position, for in the equatorial position a severe 1,3-steric interaction may develop between the methyl group and the adjacent carbonyl group[29]. The interconversion between the cis and trans forms occurs through the enamine tautomer. The imine-enamine tautomerism is solvent-dependent, too. For example at equilibrium in deuterochloroform, only 15 percent of the 9-amino-6-methyltetrahydropyridopyrimidinecarboxylate exists as the enamine tautomer, whereas in a more polar solvent such as hexadeuterodimethyl sulfoxide the proportion of the enamine form is 75 percent.

DMSO-d$_6$

R	imine	enamine	
H	75 %	25 %	Scheme 20.
CH$_3$	30 %	70 %	

The presence of a methyl group at position 6 also increases the content of the enamine tautomer (Scheme 20). In hexadeuterodimethyl sulfoxide, the predominant form of 9-anilino-3-carboxylic acid is the imine tautomer, while for the 6-methyl derivative it is the enamine form. The role of the methyl group at position 6 could be analyzed by a comparison of the X-ray data on 6-methyl-6,7,8,9-tetrahydropyridopyrimidine-3-carboxylate[30] and 1,6-dimethyl-1,6,7,8-tetrahydropyridopyrimidine-3-carboxylic acid[31].

The presence of a methyl group at position 6 is the source of two unfavourable interactions (Scheme 21). One of them is the earlier-mentioned 1,3-allylic strain between C(4)=O and C(6)-Me. This strain between the groups is relieved when the methyl group assumes the quasi-axial position. The other factor is a 1,3-diaxial interaction between the quasiaxial methyl group and the axial hydrogen at position 8.

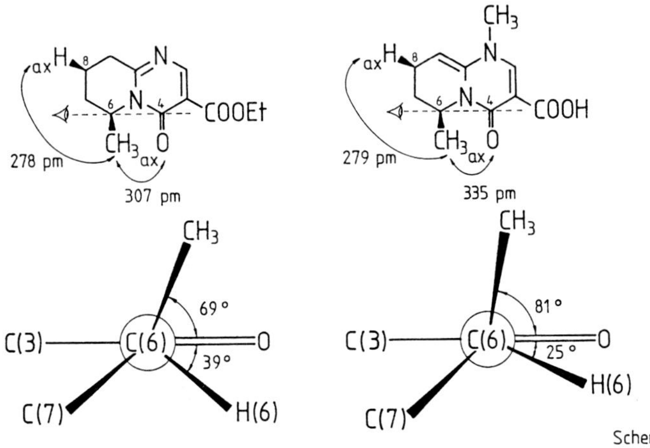

Scheme 21.

While the distances between the quasi-axial methyl group and the axial H(8) are similar in the 6,7,8,9- and 1,6,7,8-tetrahydro forms, the distance between the methyl group and the carbonyl oxygen is significantly longer in the enamine form than in the imine form. Or, if we consider the torsional angle of the methyl and carbonyl groups around the C(4) - C(6) axis in the 6,7,8,9- and 1,6,7,8-tetrahydro forms, we observe that in the former the torsional angle is 69 degrees, whereas in the 1,6,7,8-tetrahydro derivative it is 81 degrees. These data suggest that the higher proportion of the enamine form of 9-amino-6-methyltetrahydropyridopyrimidine may be attributed to further relief of the 1,3-allylic strain which is still present in the imine tautomer.

The 9-aminotetrahydropyridopyrimidines could be oxidized to 9-aminodihydro derivatives when air was bubbled through their solutions[27]. The autooxidation of 9-aminotetrahydropyridopyrimidines also proceeds in the dark and without any catalyst, which excludes the involvement of excited-state singlet oxygen and suggests a radical mechanism. Since the oxidation was much faster for those compounds where a higher proportion of the enamine tautomer could be detected, we suggested the following mechanism for the autooxidation of 9-aminotetrahydropyridopyrimidines[27].

The autooxidation of 9-aminotetrahydropyridopyrimidines starts with an electron abstraction from the enamine tautomer (Scheme 22). This is followed by an attack of triplet oxygen on the carbon C(9) of the pyridopyrimidine, and then the uptake of one electron and formation of a hydroperoxide, from which the dihydropyridopyrimidine is obtained by the elimination of hydrogen peroxide.

In contrast to the 6-methyl derivative, an unexpected reaction

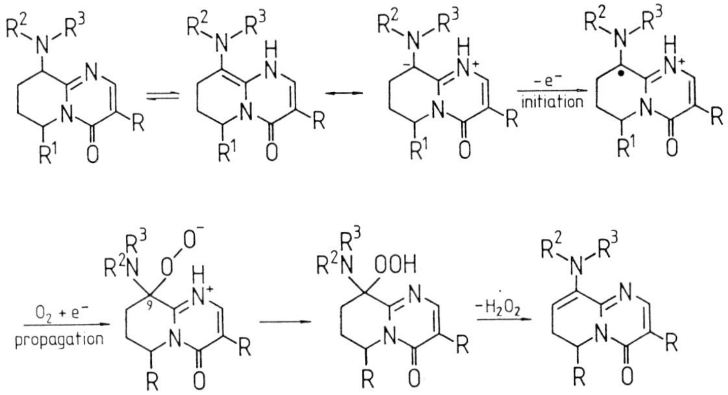

Scheme 22.

took place when 9-bromotetrahydropyridopyrimidine-3-carboxylate was reacted with N-methylaniline in boiling ethanol under an inert atmosphere (Scheme 23). The structure of the product could not be characterized as either the 9-anilino-6,7,8,9- or the -1,6,7,8-tetrahydro form, but the overall formula was identical to that of 9-anilinotetrahydropyridopyrimidine.

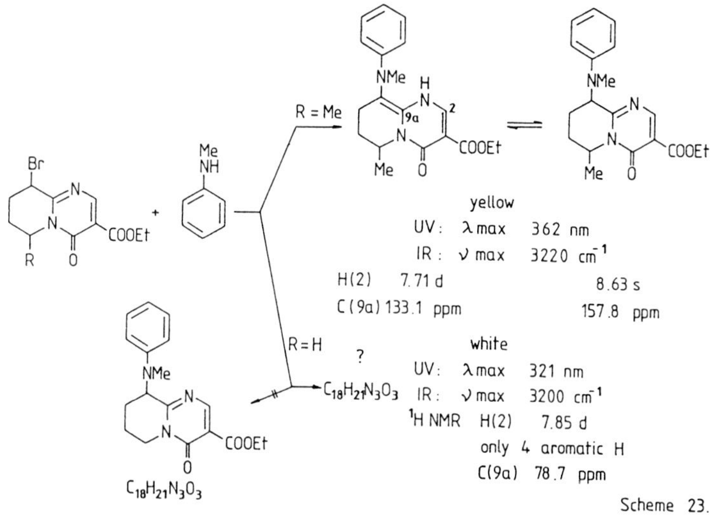

Scheme 23.

In the ^{1}H NMR spectra, the H(2) signal appeared as a doublet, which indicated that N(1) bears a hydrogen atom, but, at the same time the UV spectra excluded the presence of the 1,6,7,8-tetrahydropyridopyrimidinone chromophore. In the ^{13}C NMR spectra, the chemical shift of carbon C(9a) also appeared in an unexpected region. In the ^{1}H NMR spectra, only four aromatic protons could be detected, and the pattern of these protons indicated the presence of an orthosubstituted phenyl ring. On the basis of these data, we supposed that the ortho position of the phenyl ring and the C(9a) carbon were connected by a valence bond, and the ortho hydrogen had migrated to N(1).

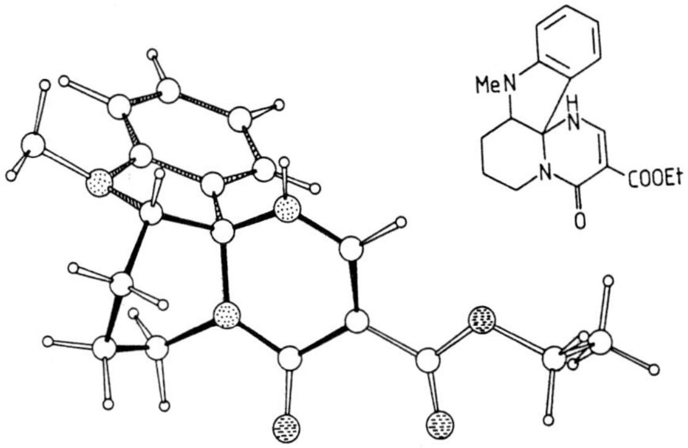

Scheme 24.

This structure was fully justified by X-ray investigations (Scheme 24). The ring junction of the pyrimidine and piperidine rings is trans, while that of the piperidine and pyrrolidine rings is cis. The piperidine ring adopts a twisted conformation.

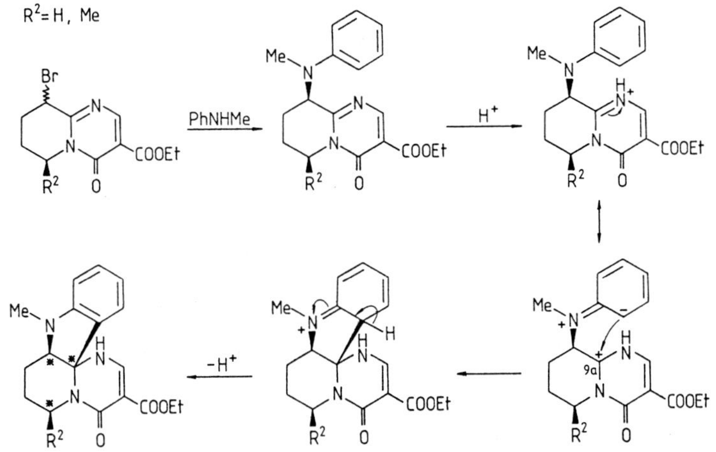

Scheme 25.

For the formation of this new tetracyclic derivative, we suggested the following mechanism: in the first step, the bromine is replaced by N-methylaniline; then, after protonation of the nitrogen at position 1, a new bond developed between the electron-deficient C(9a) and the relatively electron-rich ortho carbon of the phenyl ring (Scheme 25). Finally, aromatization of the phenyl ring takes place through the elimination of the ortho hydrogen to give the tetracyclic heterocycle.

In other words, the formation of the tetracyclic ring system is a proton-catalysed intramolecular nucleophilic addition to the C(9a)=N(1) double bond. If this is true, we should also obtain this type of compound from the 6-methyl-9-bromo derivative with N-methylaniline if the reaction is carried out under acidic conditions. In fact, when the reaction was carried out in ethanolic hydrogen chloride, we obtained the expected tetracyclic product.

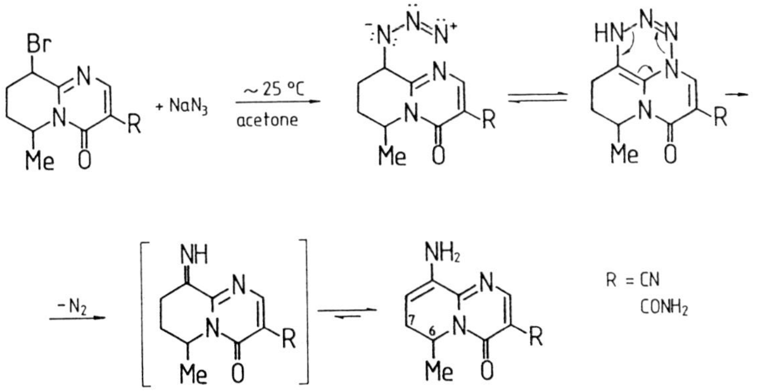

Scheme 26.

Unsubstituted 9-amino-6,7-dihydropyridopyrimidines were obtained when 9-bromopyridopyrimidines were reacted with sodium azide in acetone at ambient temperature[32]. The 9-aminoderivatives are probably formed from the ring tautomer through the elimination of a nitrogen molecule (Scheme 26). Another possibility to explain the reaction mechanism is the formation of a nitrene intermediate, but this is not very probable, if we take into consideration the very mild reaction conditions.

The 9-bromopyridopyrimidines readily react with sodium nitrite to give 9-nitropyridopyrimidines, which exist as 1,6,7,8-tetrahydro tautomers[33] (Scheme 27).

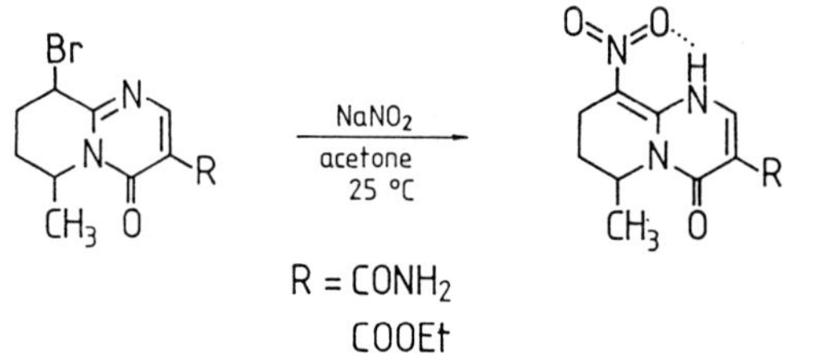

Scheme 27.

The active methylene group can be formylated with the dimethylformamide-phosphoryl chloride complex[18,19,34-38]. If an electron-withdrawing group is present at position 3, then the products are 9-dimethylaminomethylene derivatives; in the other case, the hydrolysed 9-formyl compounds can be isolated[34] (Scheme 28). The formyl derivatives exist as 1,6,7,8-tetrahydro tautomers. If the reaction mixture is treated with ethanol, or a dimethylaminomethylene derivative is reacted with ethanolic hydrogen chloride, 9-ethoxymethylene derivatives can be obtained[34].

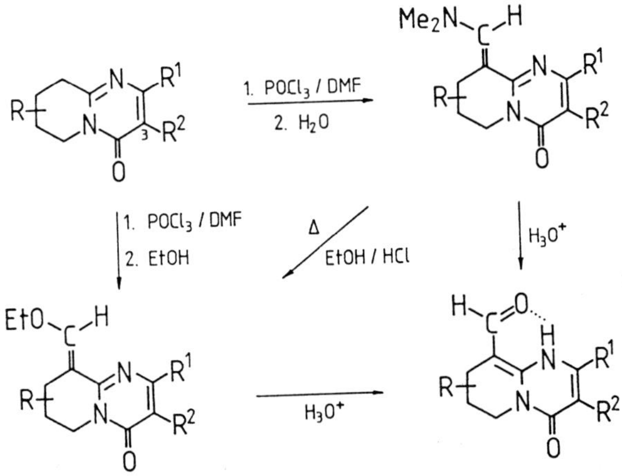

Scheme 28

Starting from 3-unsubstituted pyridopyrimidines, reaction takes place not only at position 9, but, also at position 3 (Scheme 29). Under mild conditions, at room temperature, 9-formyl derivatives were obtained, but at higher temperature the products were 3,9-diformyl derivatives[34].

We investigated some reactions of formyl derivatives (Scheme 30). The 9-formyl group could be hydrolysed under acidic or basic conditions, to give the starting 9-unsubstituted pyrido-

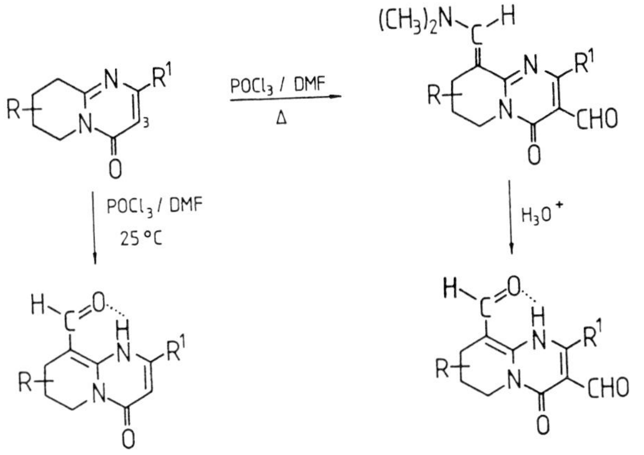

Scheme 29.

pyrimidines[34]. The formyl group readily reacted with amines, and 9-aminomethylene derivatives were obtained[21,37]. In the Japp-Klingeman reaction[39], both 9-formyl- and 9-dimethylaminomethylene derivatives afforded 9-hydrazono derivatives with aryldiazonium chlorides[18,20,37].

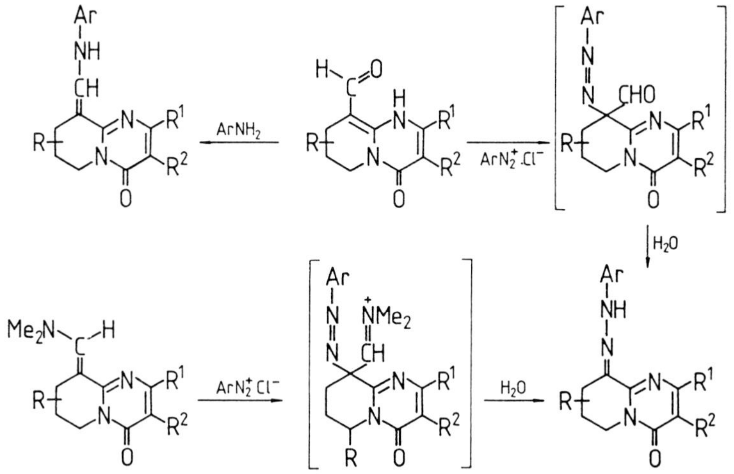

Scheme 30.

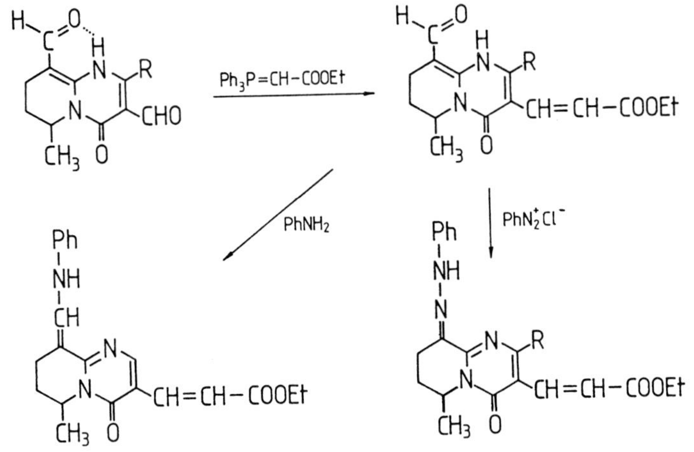

Scheme 31.

From 3,9-diformyl derivatives, 9-anilinomethylene and 9-hydrazonopyridopyrimidine-3-acrylates could be prepared[37]. In the first step the formyl group at position 3 was converted into acrylate in the Wittig reaction with (carbethoxymethylene)triphenylphosphorane (Scheme 31). Both *cis* and *trans* acrylates were formed, but only the major *trans* isomer was isolated.

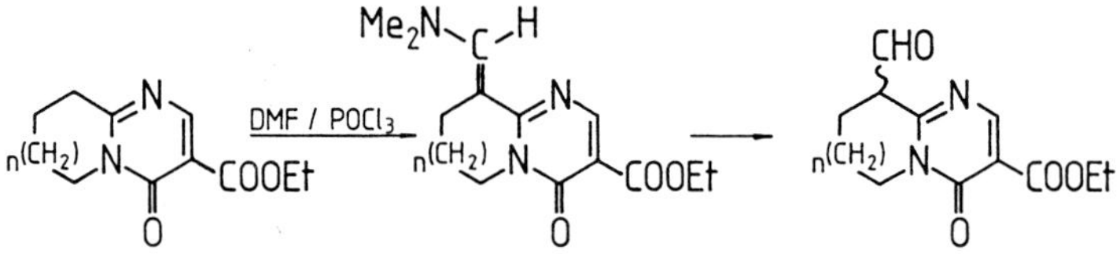

Reaction conditions

n	Temperature °C	Time h	Yield %	m.p. °C
1	25	2	80	151 - 153
0	60	3	75	183
2	85	2	82	138
3	100	2	70	144
4	100	3	65	132

Scheme 32.

The Vielsmeier-Haack formylation was extended to the piperidine ring homologues of tetrahydropyridopyrimidinone-3-carboxylates[36]. The higher homologues had to be reacted at higher temperatures

(Scheme 32). The investigations of the structures of the formylated products led to some interesting results[34,36,38,40-42].

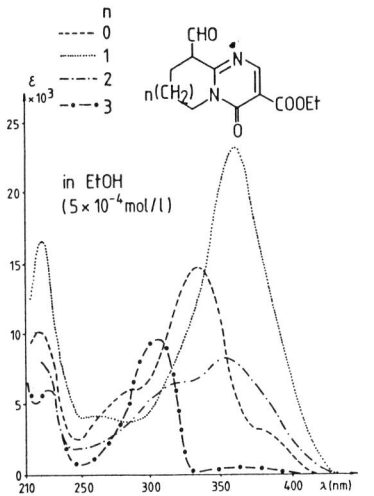

Scheme 33.

For example, while the UV spectra of the starting nitrogen bridgehead derivatives were almost identical[10], those of the formylated products were substantially different[36] (Scheme 33). These results may be interpreted so that the formyl derivatives exist in different tautomeric forms or in different proportions of the tautomeric forms.

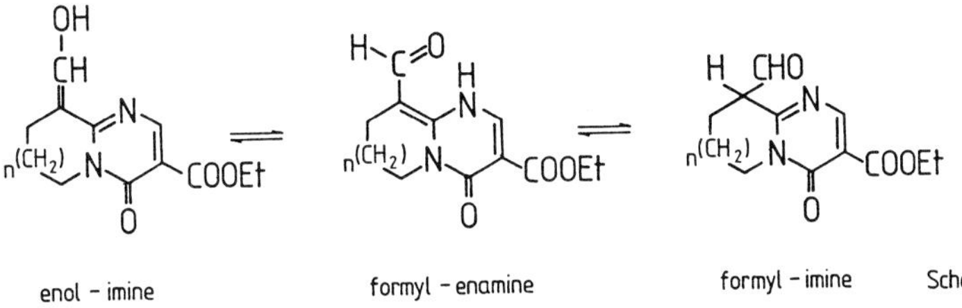

Scheme 34.

After very detailed UV, ^{1}H, ^{13}C and ^{15}N NMR investigations, we came to the conclusion that the homologues of the formyl derivatives exhibit a ring-size dependent tautomerism, where the following tautomeric forms must be considered[36,38,42] (Scheme 34).

The enol-imine tautomer contains a hydroxymethylene group, the formyl-enamine tautomer contains a <u>conjugated</u> formyl group, and the formyl-imine tautomer contains a <u>non-conjugated</u> formyl group.

The UV spectra of the formylated derivatives, except that of pyridopyrimidine, were solvent-dependent and sometimes concentration-dependent, too[36]. For example, the five-membered homologues, which exist in the enol-imine tautomeric form, are present in chloroform mainly as the Z geometric isomer, but in dioxane as the E geometric isomer (Scheme 35). The UV spectra in dioxane are very similar to that of the ethoxymethylene derivative, which exists only as the E geometric isomer. The next picture shows that the five-membered derivative is mainly present in concentrated solution as the E isomer, but in dilute solution as the Z isomer.

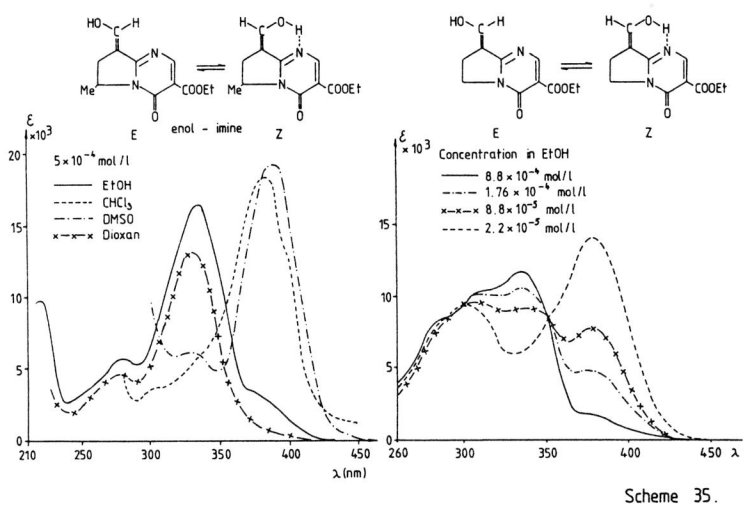

Scheme 35.

In studies of this ring-size dependent tautomerism by NMR, it must be borne in mind that the prototropic processes taking place between two heteroatoms are fast in relation to the NMR time-scale, and only time-averaged signals can be observed. In our case this is the situation as regards the enol-imine and formyl-enamine tautomers. In contrast, prototropic processes taking place between a carbon atom (formyl-imine tautomers) and a heteroatom (enol-imine or formyl-enamine tautomers) are usually slow, and thus the signals of the individual tautomers can be distinguished.

In fact the ^{1}H NMR spectra of the pyridopyrimidines contain only one set of signals, while the spectra of higher homologues contain two sets of signals with different intensities.

Scheme 36 shows some characteristic ^{1}H NMR data. The proportion of the formyl-imine tautomer could be estimated, for example, from the intensities of the two formyl protons.

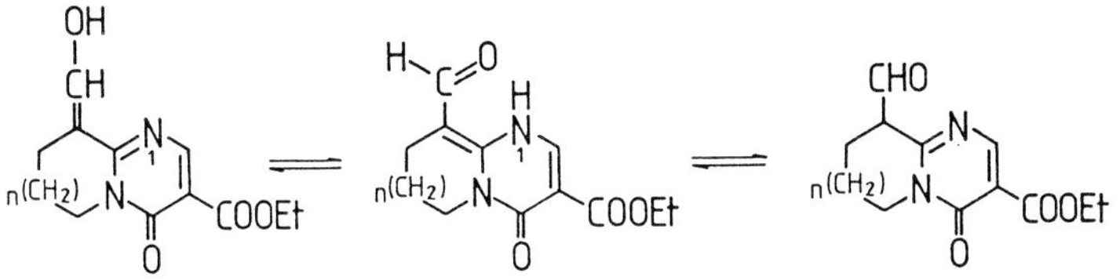

Scheme 36

n	δ CHO	H-2	$^1J_{15N,H}$	in CDCl	δ CHO	H-2
	enol-imine				formyl-imine	
0	7.31	8.48 s				
1	8.78 s	8.35 d	81.2 Hz			
2	8.72 s	8.18 d			9.90 d	8.49 s
3	8.70 s	8.26 d			9.88 d	8.50 s
4	8.63 s	8.29 d			9.78 d	8.63 s

$^1J_{15N-H} \approx 92$ Hz in enamine

formyl-enamine % = $\dfrac{81.2\ Hz - 0\ Hz}{92\ Hz - 0\ Hz} \cdot 100 = \sim 87\%$

Scheme 36.

To estimate the proportions of the enol-imine and formyl-enamine tautomers, nitrogen-15 NMR was applied[40,42]. We used either the nitrogen-15 - proton coupling constants, or the nitrogen-15 chemical shifts. In the latter case, beside the formyl derivatives, fixed forms of the enol-imine and formyl-enamine tautomers labelled with ^{15}N at position 1 were applied. For example, a value of 92 Hz has been reported for the nitrogen-15 - proton coupling constant at enamines[43]. The corresponding value is naturally zero for imines. In the six-membered homologues we

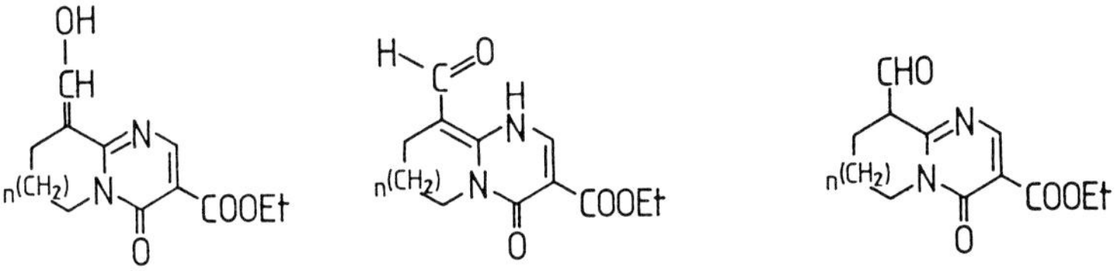

n	enol-imine	formyl-enamine	formyl-imine
0	~ 100 %		
1	~ 11 - 15 %	~ 85 - 89 %	
2	~ 15 %	~ 75 %	10 %
3	~ 10 %	~ 60 %	30 %
4	~ 3 %	~ 17 %	80 %

in CDCl$_3$ (10^{-3} mol/l)

Scheme 37.

measured 81.2 Hz. Accordingly 81.2 must be divided by 92 and multiplied by 100, which gives roughly 87, that is, the equilibrium mixture of formylpyridopyrimidine consists of 87 per cent formyl-enamine and 13 per cent enol-imine plus formyl-imine.

Scheme 37 summarizes the results of our investigations in deuterochloroform solution. The five-membered homologue exists practically only as the enol-imine tautomer, but the six-membered derivative as a mixture of enol-imine and formyl-enamine tautomers. For the higher homologues, the presence of all three tautomers could be detected in the mixture, and for the nine-membered homologue the formyl-imine form is already the predominant tautomer.

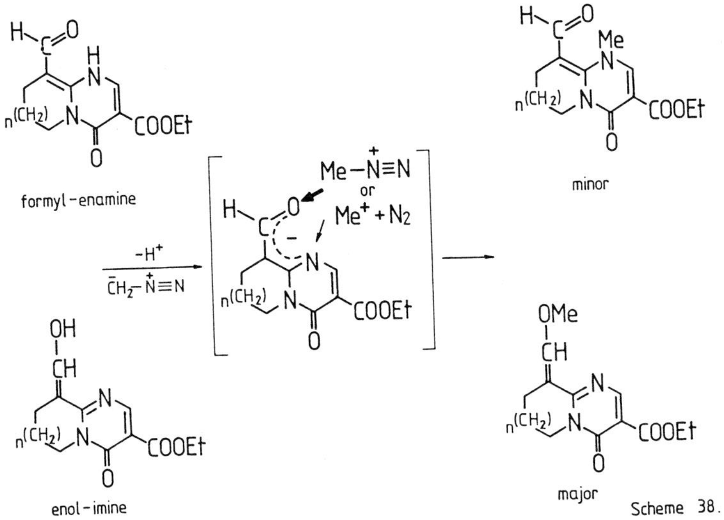

Scheme 38.

We wondered whether the different tautomeric forms of the formyl derivatives could be trapped in a reaction. For this purpose we studied the reactions of the formyl derivatives with diazomethane in chloroform at room temperature (Scheme 38).

We can expect that, in the case of enol-imine and formyl-enamine tautomers, after a fast proton transfer, the protonated diazomethane or methyl cation will attack on the harder oxygen atom to give the methoxymethylene derivative as the major product. In the case of the formyl-imine tautomer (non-conjugated formyl group), diazomethane is expected to react with the electrophilic carbon of the formyl group and, after the loss of a nitrogen molecule, a tricyclic derivative will form in a ring-closure reaction between the carbo cation and N(1), after a proton migration (Scheme 39). In practice we could isolate the dehydrated tricyclic derivatives.

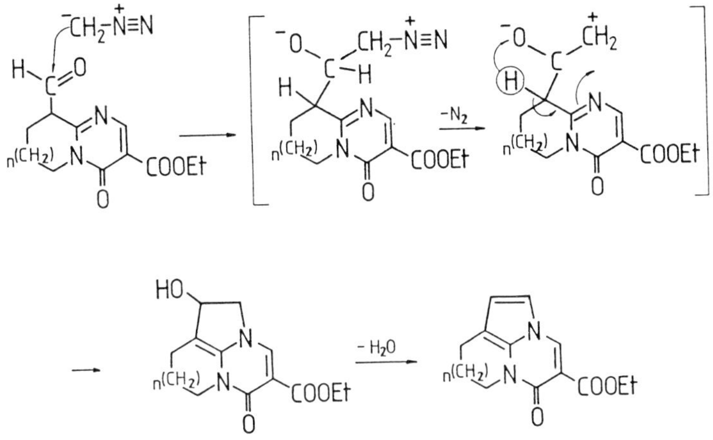

Scheme 39.

Scheme 40 summarizes the results of the reactions with diazomethane. From the five- and six-membered derivatives we obtained O-methylated products, together with minor quantities of N-methylated compounds.

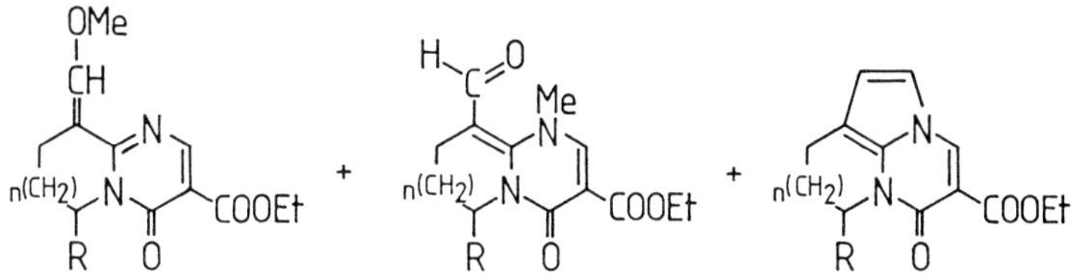

n	R			
0	Me	54.9 %	8.4 %	
1	Me	54.0 %	4.0 %	
2	H	50.0 %		5.0 %
3	H			39.0 %
4	H			62.0 %

Scheme 40.

The seven-membered derivative gave mainly the O-methylated product, with 5 per cent of the tricyclic compound formed from the formyl-imine tautomer, and in the case of the eight- and nine-membered homologues we could isolate only the tricyclic derivatives.

To modify the structures of the nitrogen bridgehead derivatives, we used the clay-fen reagent, too, which is a clay-supported ferric nitrate[44]. Clay-fen can be regarded as a reagent capable of oxidizing and/or nitrating a substrate under mild conditions[45].

For example, 9-isonitrosopyridopyrimidine was stirred in dichloromethane in the presence of clay-fen at room temperature. After 6 hours, the clay-fen was filtered off, the reaction mixture was evaporated, and the residue was chromatographed. In this way we obtained 9-nitropyridopyrimidine, which, as we saw earlier, was also prepared in the reaction of 9-bromopyridopyrimidine with sodium nitrite[33] (Scheme 41).

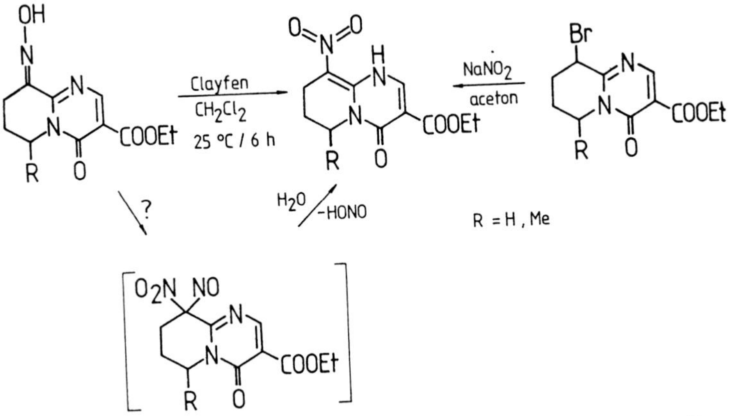

Scheme 41.

As concerns the formation of the 9-nitro derivatives, we considered two possibilities: the oxime group was oxidized directly to the nitro group, or in the first step the 9-nitro-9-nitroso intermediate was formed by nitration, and this intermediate was then hydrolysed to the 9-nitro derivative. To choose between these two possibilities, we carried out the following experiments: 9-formylpyridopyrimidine was reacted with clay-fen under similar conditions (Scheme 42). In this case we also obtained the 9-nitro derivative, which was probably formed from the 9-nitro-9-formyl intermediate. In another experiment, we used the isonitroso derivative which was labelled with N-15 to 50% enrichment. If nitration takes place followed by hydrolysis, then the 9-nitro derivative formed will not contain N-15 while if oxidation of the isonitroso group occurs, the product must contain N-15. We experienced that the product does contain N-15 to 50% enrichment.

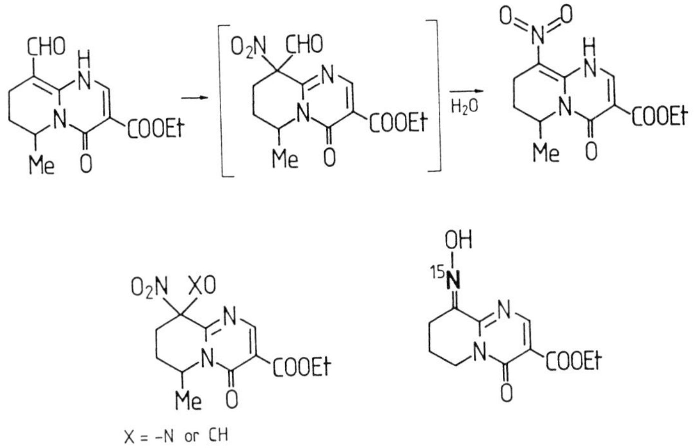

Scheme 42.

That is, in the case of isonitroso derivatives clay-fen can be regarded as an oxidation reagent, while in the case of formyl derivatives it behaves as a nitration agent.

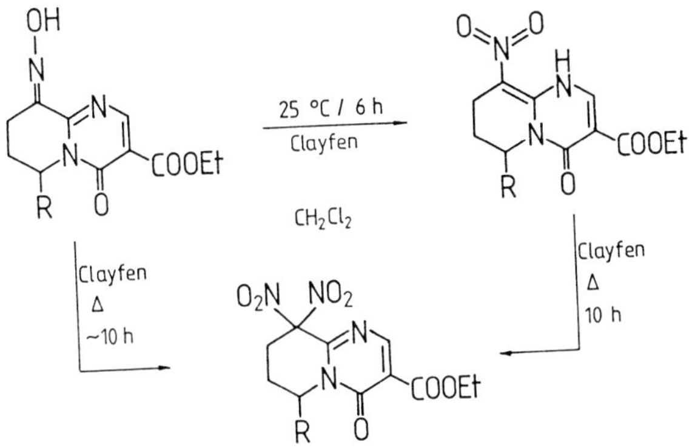

Scheme 43.

If we reacted the isonitroso derivatives with clay-fen under forced conditions, at reflux temperature for 10 h, we obtained dinitropyridopyrimidines (Scheme 43). These products were also obtained from 9-nitro derivatives. In this case, clay-fen reacted as an oxidation and a nitration reagent.

An unexpected reaction was observed when the seven-membered formylazepinopyrimidine was reacted with clay-fen (Scheme 44). When the reaction was carried out at a one to one molar ratio, we obtained dinitroazepinopyrimidine, but when only half a mol of clay-fen was used for one mol of formyl derivative, the product was 9-nitropyridopyrimidine. That is, besides the nitration, ring contraction also occurred. At present we have no idea about the reaction mechanism.

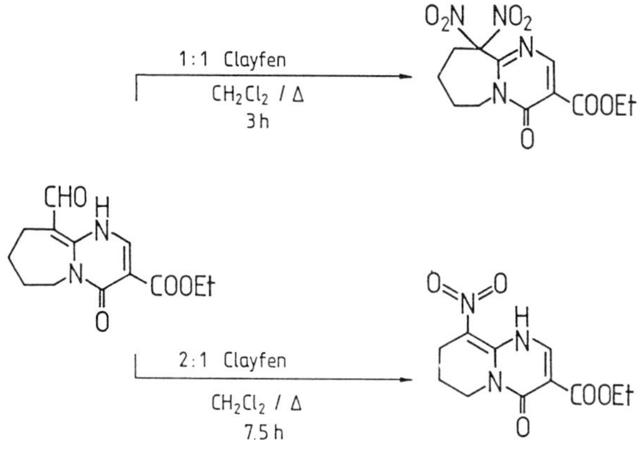

Scheme 44.

Since many of the 9-substituted pyridopyrimidines mentioned earlier exhibited high antiallergic activities, we also investigated their reactivity in different reactions.

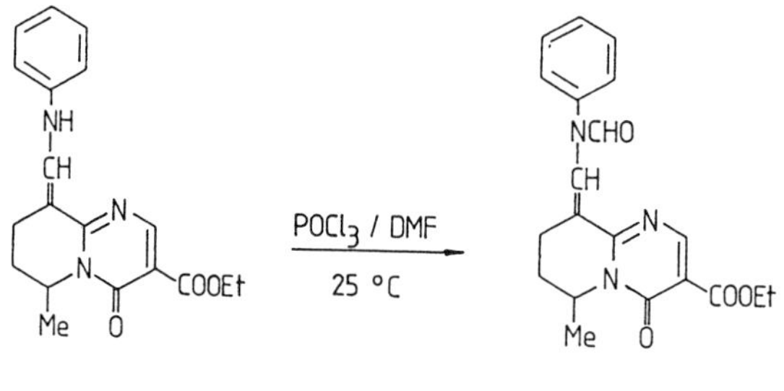

Scheme 45.

For example, the reactions of 9-substituted pyridopyrimidines with the Vilsmeier-Haack reagent were investigated. The reaction of the 9-phenylaminomethylene derivative with dimethylformamide and phosphoryl chloride gave an N-formylated product (Scheme 45).

At the end of the lecture, we shall summarize very briefly the results of structure-activity relationship studies. We obtained antiallergic derivatives with high activities among the 9-oxime[9], 9-amino[9], 9-arylaminomethylene[37] and 9-arylhydrazono derivatives[15,16,37,46]. In each series we found similar structure-activity relationships. In all series only the 3-carboxylic acids exhibit good antiallergic properties (Scheme 46).

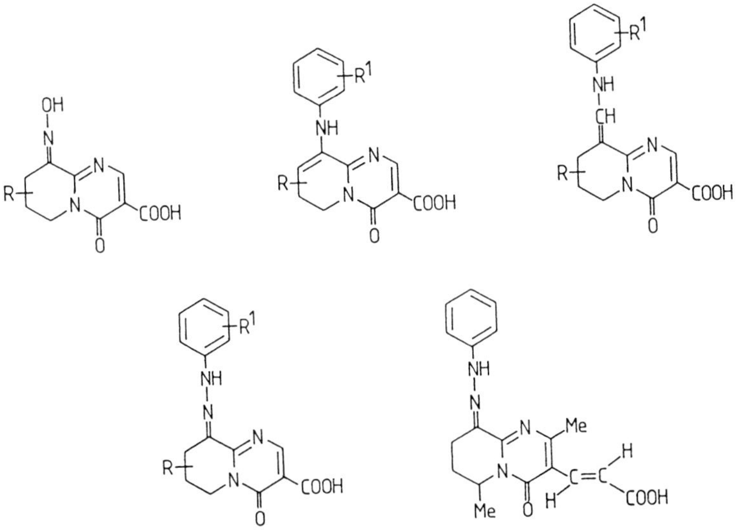

Scheme 46.

If the OH group was substituted, or the amino groups were disubstituted, the activities were lost. The activity was also lost if a substitutent was present in position 2. However, in the latter

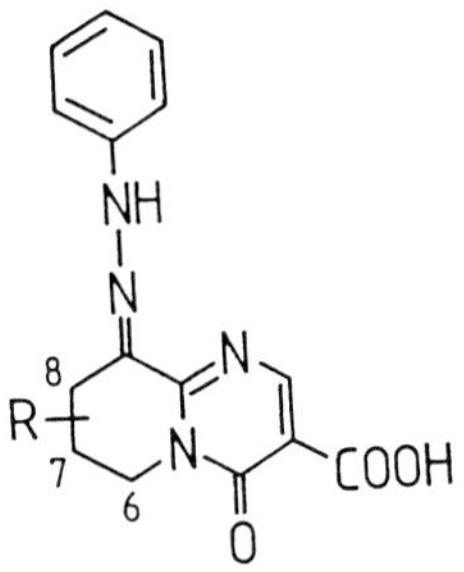

PCA test (rat) i.v.

	ID$_{50}$ µMol/kg	R	ID$_{50}$ µMol/kg
Intal	1.0	(RS)(±)6-Me	0.6
R = H	31.6	(S)(+)6-Me	0.3
R = 6-Me	0.6	(R)(−)6-Me	54.8
R = 7-Me	5.7		
R = 8-Me	32.0		

Scheme 47.

case, if the pyridopyrimidine ring and carboxyl group were linked by a trans vinyl moiety, high antiallergic activity again appeared[37].

In all series the most active derivatives are the 6-methyl derivatives, and the antiallergic activity resides in the optically active form containing the S absolute configuration in position 6 (Scheme 47).

ACKNOWLEDGEMENT:

An account of this work would not be complete without due mention of the enthusiastic and hard-working colleagues who made it possible. They are: L. Vasvári-Debreczy, Á. Horváth, M. Balogh, J. Kökösi, G. Horváth, K. Simon, I. Bitter, G. Tóth, M. Kajtár, C. DeVos (Belgium), F. Dessy (Belgium), L. Rodriguez (Belgium), T. Breining (Canada), Abdirizak Osman Hassan (Somali), J. Sessi (Nigeria).

REFERENCES:

1. I. Hermecz and Z. Mészáros, Med. Res. Rev., 1988, **8**, 203.
2. J. Knoll, Z. Mészáros, P. Szentmiklósi, and S. Fürst, Arzneim.-Forsch., 1971, **21**, 717.
3. Z. Mészáros, J. Knoll, P. Szentmiklósi, Á. Dávid, G. Horváth, and I. Hermecz, Arzneim.-Forsch., 1972, **22**, 815.
4. J. Knoll, K. Gyires, and Z. Mészáros, Arzneim.-Forsch., 1979, **29**, 763.
5. I. Hermecz, M. Kajtár, K. Simon, T. Breining, R. P. Surján, G. Tóth, and Z. Mészáros, J. Org. Chem., 1985, **50**, 2918.
6. a. Z. Mészáros, J. Knoll, P. Szentmiklósi, I. Hermecz, Á. Horváth, S. Virág, L. Vasvári-Debreczy, and Á. Dávid, Ger. Patent 2,653,257; Chem. Abstr., 1977, **87**, 135383.
   b. J. Bédi, Gy. Blaskó, and L. A. Pálos, Arzneim.-Forsch., 1979, **29**, 1405.
7. G. Náray-Szabó, I. Hermecz, and Z. Mészáros, J. C. S. Perkin Trans. 1, 1974, 1753.
8. I. Hermecz, Z. Mészáros, L. Vasvári-Debreczy, Á. Horváth, S. Virág, and J. Sipos, Arzneim.-Forsch., 1979, **29**, 1833.
9. I. Hermecz, T. Breining, L. Vasvári-Debreczy, Á. Horváth, Z. Mészáros, I. Bitter, C. DeVos, and L. Rodriguez, J. Med. Chem., 1983, **26**, 1494.
10. J. Kökösi, I. Hermecz, Gy. Szász, Z. Mészáros, G. Tóth, and M. Csákvári-Pongor, J. Heterocyclic Chem., 1982, **19**, 909.
11. I. Hermecz, T. Breining, Z. Mészáros, G. Tóth, and I. Bitter, Heterocycles, 1980, **14**, 1953.
12. M. Kajtár, J. Kajtár, I. Hermecz, T. Breining, and Z. Mészáros, J. Heterocyclic Chem., 1987, **24**, 393.
13. O. S. Tee and S. Banerjee, J. Org. Chem., 1979, **44**, 3256.
14. G. Tóth, Á. Szöllősy, A. Almásy, B. Podányi, I. Hermecz,

T. Breining, and Z. Mészáros, Org. Magn. Reson., 1983, **21**, 687.
15. I. Hermecz, T. Breining, Z. Mészáros, Á. Horváth, L. Vasvári-Debreczy, F. Dessy, C. DeVos, and L. Rodriguez, J. Med. Chem., 1982, **25**, 1140.
16. I. Hermecz, T. Breining, Z. Mészáros, J. Kökösi, L. Mészáros, F. Dessy, and C. DeVos, J. Med. Chem., 1983, **26**, 1126.
17. J. Kökösi, I. Hermecz, Gy. Szász, and Z. Mészáros, Tetrahedron L., 1981, **22**, 4861.
18. J. Kökösi, I. Hermecz, B. Podányi, Gy. Szász, and Z. Mészáros, J. Heterocyclic Chem., 1984, **21**, 1301.
19. I. Hermecz, J. Kökösi, Á. Horváth, B. Podányi, L. Vasvári-Debreczy, Gy. Szász, and Z. Mészáros, J. Heterocyclic Chem., 1987, **24**, 1045.
20. J. Kökösi, I. Hermecz, B. Podányi, Gy. Szász, and Z. Mészáros, J. Heterocyclic Chem., 1985, **22**, 1373.
21. G. Tóth, B. Podányi, I. Hermecz, Á. Horváth, G. Horváth, and Z. Mészáros, J. Chem. Res. (S), 1983, 161 & J. Chem. Res. (M), 1983, 1721.
22. H. O. Kalinovski and H. Kessler, 'Topics in Stereochemistry' ed: N. L. Allinger, E. L. Eliel, S. H. Wilen & Sons, Wiley, New York, 1973, Vol. 3, p. 295.
23. J. Sandström, 'Topics in Stereochemistry' ed: N. L. Allinger E. L. Eliel, S. H. Wilen, Wiley & Sons, New York, 1983, Vol. 14, p. 83.
24. B. Robinson, 'The Fischer Indol Synthesis' Wiley & Sons, New York, 1982.
25. G. M. Robinson and R. Robinson, J. Chem. Soc., 1918, **113**, 639.
26. T. Breining, I. Hermecz, B. Podányi, and J. Sessi, J. Heterocyclic Chem., 1985, **22**, 1253.
27. I. Hermecz, R. P. Surján, T. Breining, K. Simon, G. Horváth, Z. Mészáros, M. Kajtár, and G. Tóth, J. C. S. Perkin Trans. 2, 1983, 1413.
28. T. Breining, I. Hermecz, B. Podányi, Z. Mészáros, and G. Tóth, J. S. S. Perkin Trans. 1, 1985, 1015.
29. F. Johnson, Chem. Rev., 1968, **68**, 375.
30. K. Sasvári and K. Simon, Acta Crystallogr., 1973, **B29**, 1245.
31. K. Simon, Z. Mészáros, and K. Sasvári, Acta Crystallogr., 1975, **B31**, 1701.
32. I. Bitter, G. Tóth, Á. Szöllősy, I. Hermecz, and Z. Mészáros, Tetrahedron L., 1985, **26**, 3521.
33. I. Bitter, G. Tóth, I. Hermecz, and Z. Mészáros, Heterocycles, 1987, **26**, 869.
34. Á. Horváht, I. Hermecz, L. Vasvári-Debreczy, K. Simon, M. Pongor-Csákvári, and G. Tóth, J. C. S. Perkin Trans. 1, 1983, 369.
35. I. Hermecz, I. Bitter, Á. Horváht, G. Tóth, and Z. Mészáros, Tetrahedron L., 1979, 2557.
36. I. Hermecz, Á. Horváth, Z. Mészáros, M. Pongor-Csákvári, G. Tóth, and Á. Szöllősy, J. C. S. Perkin Trans. 2, 1985, 1873.
37. I. Hermecz, Á. Horváth, Z. Mészáros, C. DeVos, and L.

Rodriguez, J. Med. Chem., 1984, **27**, 1253.
38. Á. Horváth, I. Hermecz, M. Pongor-Csákvári, Z. Mészáros, J. Kökösi, G. Tóth, and Á. Szöllősy, J. Heterocyclic Chem., 1984, **21**, 219.
39. R. R. Phillips, Org. React., 1959, **10**, 143.
40. G. Tóth, Á. Szöllősy, Cs. Szántay, Jr., I. Hermecz, Á. Horváth, and Z. Mészáros, J. C. S. Perkin Trans. 2, 1983, 1153.
41. G. Tóth, Á. Szöllősy, Cs. Szántay, Jr., A. Almásy, I. Hermecz, Á. Horváth, L. Vasvári-Debreczy, and Z. Mészáros, J. C. S. Perkin Trans. 2, 1986, 1911.
42. G. Tóth, Á. Szöllősy, I. Hermecz, Á. Horváth, Z. Mészáros, J. C. S. Perkin Trans. 2, 1985, 1881.
43. G. O. Dudek and E. P. Dudek, J. Am. Chem. Soc., 1964, **86**, 4283.
44. M. Balogh, P. Pennetreau, I. Hermecz, and A. Gerstmans, J. Org. Chem., in press.
45. A. Cornelis and P. Laszlo, Synthesis, 1985, 909.
46. C. DeVos, F. Dessy, I. Hermecz, T. Breining, and Z. Mészáros, Int. Arch. Allergy Appl. Immunol., 1982, **67**, 362.

Heterocycles Designed for the Modelling of the Zinc Active Site of Liver Alcohol Dehydrogenase

R. M. Kellogg and B. Kaptein

DEPARMENT OF ORGANIC CHEMISTRY, UNIVERSITY OF GRONINGEN,
NIJENBORGH 16, 9747 AG GRONINGEN, THE NETHERLANDS

The zinc ion at the active site of horse liver alcohol dehydrogenase (HLAD) is tetrahedrally ligated by two thiolates from cysteine, an imidazole from histidine, and a water molecule.[1,2] This type of ligation is rare in the chemistry of Zn(II) owing to the pronounced tendency of thiolate to bridge between two zinc ions. In other words dimeric or oligomeric structures are formed (*Scheme 1*).

Scheme 1

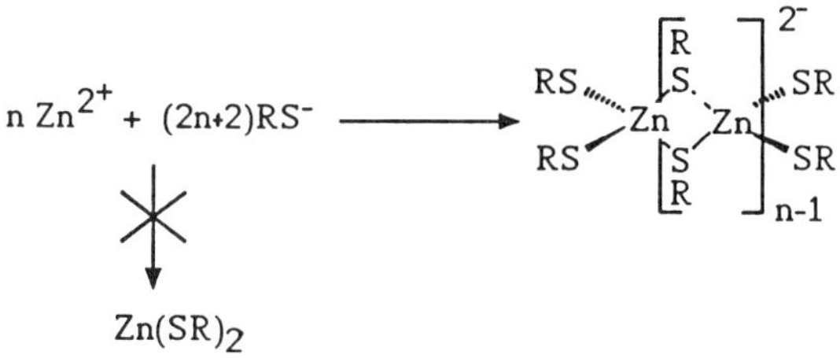

$$n\, Zn^{2+} + (2n+2)RS^- \longrightarrow \left[\begin{array}{c} R \\ RS_{\cdots\cdots}\overset{S}{\underset{S}{\diagup}}Zn\overset{\diagup}{\underset{\diagdown}{\diagdown}}Zn\overset{\cdots}{\diagdown}SR \\ RS\overset{\diagup}{\diagdown}\underset{R}{\overset{S}{\diagdown}}SR \end{array} \right]^{2-}_{n-1}$$

$$\Big\downarrow\!\!\!\!\!\!\times$$

$$Zn(SR)_2$$

1. See, for example, Eklund, H.; Plape, B.V.; Samana, J.P.; Brändén, C.-I. J. Biol. Chem., 1982, **257**, 14349.
2. Material for this article has been taken in part form Kaptein, B.; Barf G.; Kellogg, R.M.; Van Bolhuis, F.; J. Org. Chem., 1990, **55**, 1890.

We have prepared several heterocyclic ligands containing two thiolates. These were designed to ensure that zinc ions would coordinate without dimerization or oligomerization. Steric shielding was the major tactics used to prevent association. The coordination is intended to mimic the situation around zinc in HLAD.

A synthesis of **1** has been developed and this was anticipated to coordinate as indicated in *Scheme 2*. The metal ion is protected by a "roof" and "walls". Zinc complexes of **1** unfor-tunately turned out to be oligomeric. However, cobalt(II) ions complexed cleanly on a one to one basis with **1** and provided a complex with spectral properties that mimic remarkably well the visible spectra of HLAD wherein Zn(II) at the active site has been substi-tuted by Co(II). This **1** Co(II) complex exhibited moderate catalytic activity.

Scheme 2

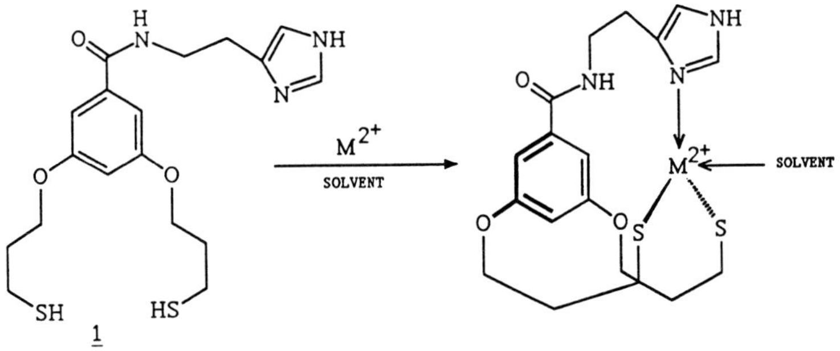

Another approach to monomeric Zn(II) complexes is via hindered heterocycles like **2** (*Scheme 3*).[3] The enormous tendency of zinc thiolates to bridge is illustrated dramatically by the dimerization of this material. Both X-ray and NMR data reveal that dimerization has occurred both in solution and in the solid state.

---

3. Holm, R.H. ; Berg, J.M.; Acc. Chem. Res., 1986, **19**, 363.

Scheme 3

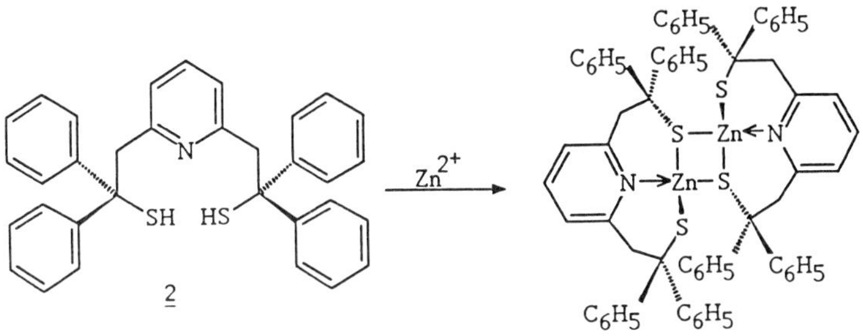

Monomeric complexes of Zn(II) could be obtained, however, when steric hindrance in **2** was better directed towards prevention of intermolecular interactions. This was achieved in **3** in which the phenyl rings are "tied together". The synthesis was carried out as illustrated in *Scheme 4*. This extremely hindered heterocycle forms a monomeric complex with Zn(II) in which the thiolates (deprotonation occurs on complexation) and pyridine nitrogen are coordinated to Zn(II). Analogous complexation also occurs with Co(II) ions. Although it has not been possible to obtain crystals suitable for X-ray determination, NMR investigations in solution clearly reveal the monomeric nature of the complexes.

Scheme 4

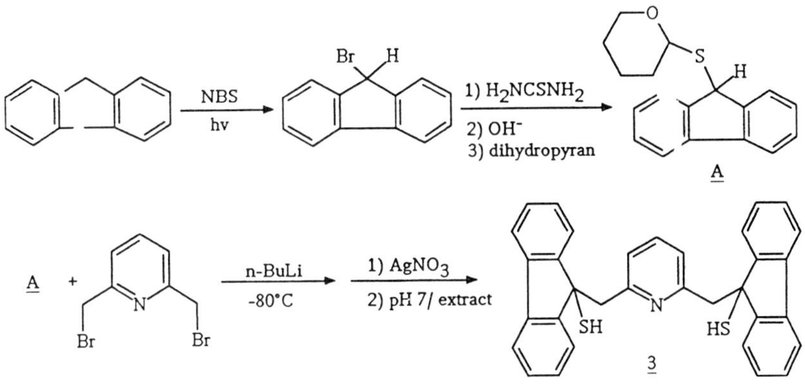

A considerable amount of extra information about zinc coordination has been obtained with **4a-d**, synthesized as illustrated in *Scheme 5*.

Scheme 5

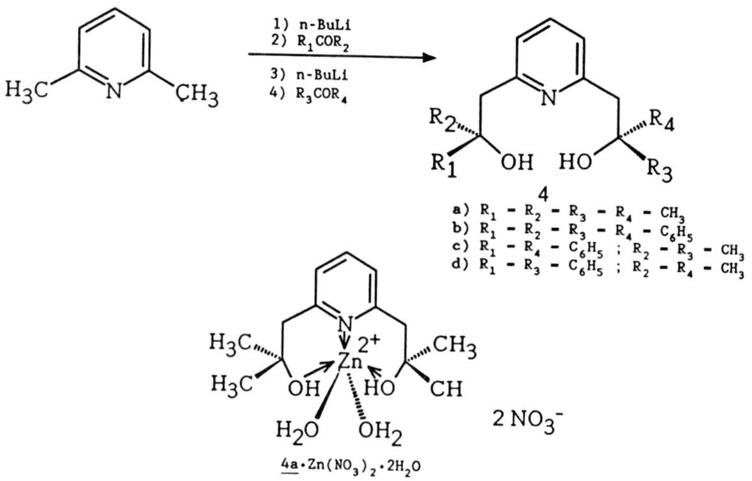

[Structure: 2,6-lutidine reacting via 1) n-BuLi, 2) R_1COR_2, 3) n-BuLi, 4) R_3COR_4 to give compound **4** with pyridine bearing two $-CR_1R_2(OH)-$ and $-CR_3R_4(OH)-$ arms]

a) $R_1 = R_2 = R_3 = R_4 = CH_3$
b) $R_1 = R_2 = R_3 = R_4 = C_6H_5$
c) $R_1 = R_4 = C_6H_5$; $R_2 = R_3 = CH_3$
d) $R_1 = R_3 = C_6H_5$; $R_2 = R_4 = CH_3$

[Structure of **4a**·Zn(NO$_3$)$_2$·2H$_2$O: pentacoordinated Zn^{2+} complex with pyridine N, two alcohol OH, and two H$_2$O ligands; 2 NO$_3^-$ counterions]

4a·Zn(NO$_3$)$_2$·2H$_2$O

Heterocycle **4a** provides a remarkable pentacoordinated complex with Zn(NO$_3$)$_2$. The pentacoordination arises from the retention of two water molecules, which coordinate zinc in addition to two alcohols (in contrast to the thiols, no deprotonation occurs) and pyridine nitrogen.

Various investigations of the catalytic activity of complexes derived from **1-4** have been made. The complexes derived from **4** bear some resemblance to catalytically active zinc in thermolysine and carboxypeptidase[4,5].

4. Matthews, B.W.; Weaver, L.H.; Kester, W.R.; J.Biol. Chem., 1974, **249**, 8030.
5. Hartsuck, J.A. ; Lipscomb, W.N.; In The Enzymes, Vol. 3; Boyer, P.D., Ed.; Academic Press: New York, 1971; pp 1-56.

Chemical Topology: Interlacing and Knotting Molecular Threads on Transition Metals

J. -P. Sauvage

LABORATOIRE DE CHIMIE ORGANO-MINÉRALE (UA 422 AU CNRS), INSTITUT DE CHIMIE, 1, RUE BLAISE PASCAL, 67000 STRASBOURG, FRANCE

Knots and interlocked rings have always exerted a real fascination on man, as witnessed by the many representations of interlaced and knotted threads found in art for centuries.

Making topologically novel molecules like catenanes (interlocked rings) and knots has been a challenge in Chemistry for decades.[1-3]

Transition metals are powerful tools in this view since they are able to bind and collect two or more coordinating molecular threads and, by further reactions, the precursor complex obtained can lead to interlaced systems. This new type of three-dimensional template effect has been developed in our group for a few years. The general principle is given in Figure 1.

The first strategy has allowed the synthesis of <u>catenates</u> (complexes) and <u>catenands</u> (free ligands containing interlocked macrocycles).[4,5] The most representative example is that of the copper (I) catenate 4^+, the ligand **5** consisting of two interlocked 30-membered rings. The sequence of reactions leading to the catenate 4^+ is given in Figure 2.

The complex 4^+ could be quantitatively demetallated using KCN, leading to the catenand **5**, as indicated in Figure 3.

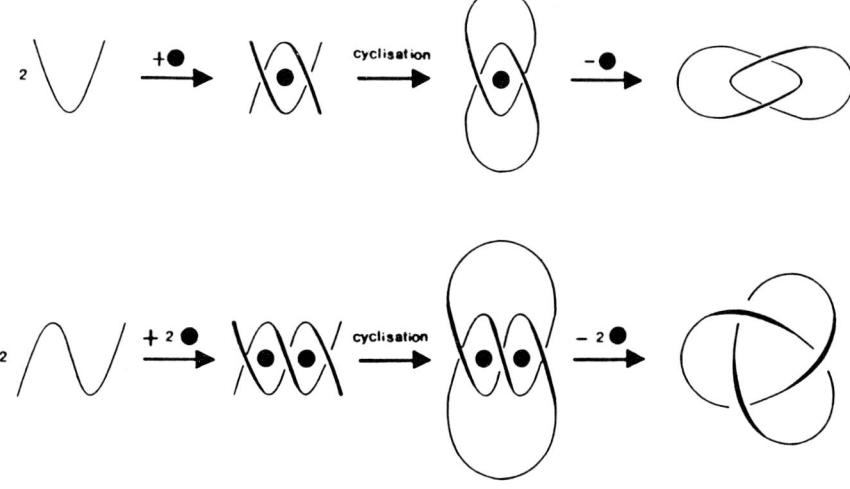

Figure 1

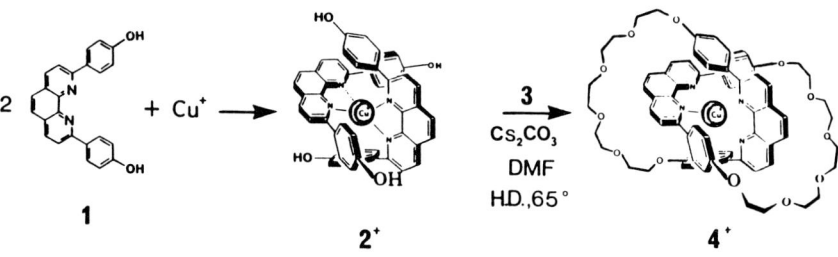

Figure 2 Two chelating fragments **1** are entwined around a copper (I) centre ; the precursor complex obtained (**2⁺**) is reacted with $ICH_2(CH_2OCH_2)_4CH_2I$ (**3**) and Cs_2CO_3 to afford (27%) the catenate **4⁺**.

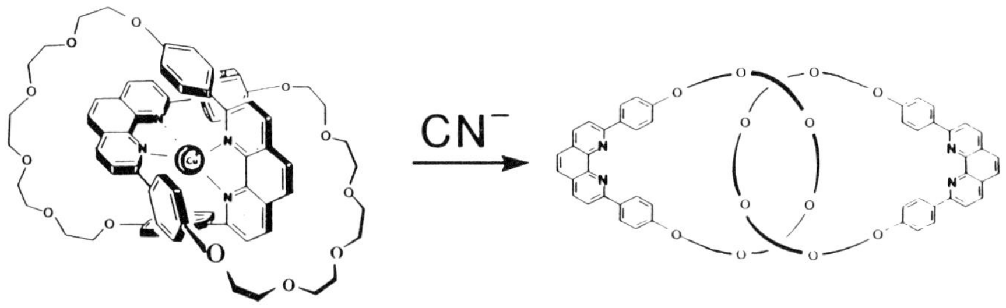

Figure 3

The particular topology of the catenands and the topography of the corresponding complexes are responsible for the very special electrochemical, photophysical and kinetic properties of the catenates. In particular, low oxidation states like Ni (I) or Cu (I) are very strongly stabilized ; divalent complexes cannot be oxidized to the +3 state.

Molecular systems containing two peripheral rings separately interlocked to a large central cycle were also prepared, as an extension of the original series.[6,7] In Figure 4 are indicated the reactions affording the di-copper (I) [3]-catenate 7^{2+}.

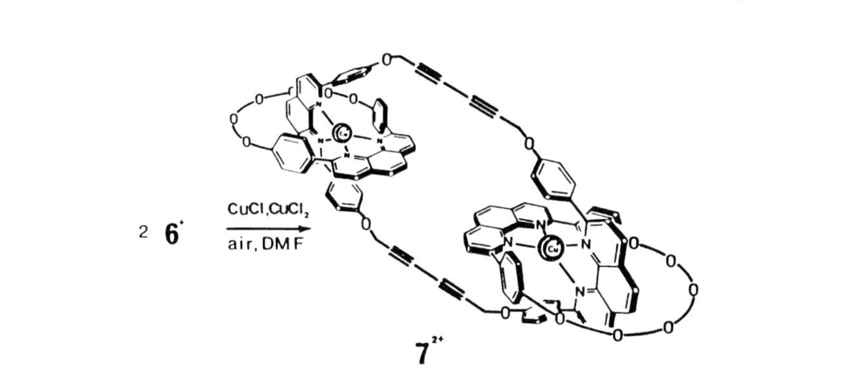

The procedure being really preparative (58% yield for the above sequence of reactions), a complete study of 7^{2+} and analogous compounds of the same family could be carried out.[8]

Recently, a synthetic molecular trefoil knot was made.[9] The strategy is based on the use of two metal centres (Cu (I)) able to gather and interlace properly two molecular threads. After cyclisation of the precursor di-nuclear double helix, a knotted complex is obtained (second strategy of Figure 1). The synthetic approach is given in Figure 5. The di-copper (I) trefoil knot 9^{2+} is obtained in 3% yield (!) from the starting bis-chelate molecular thread.

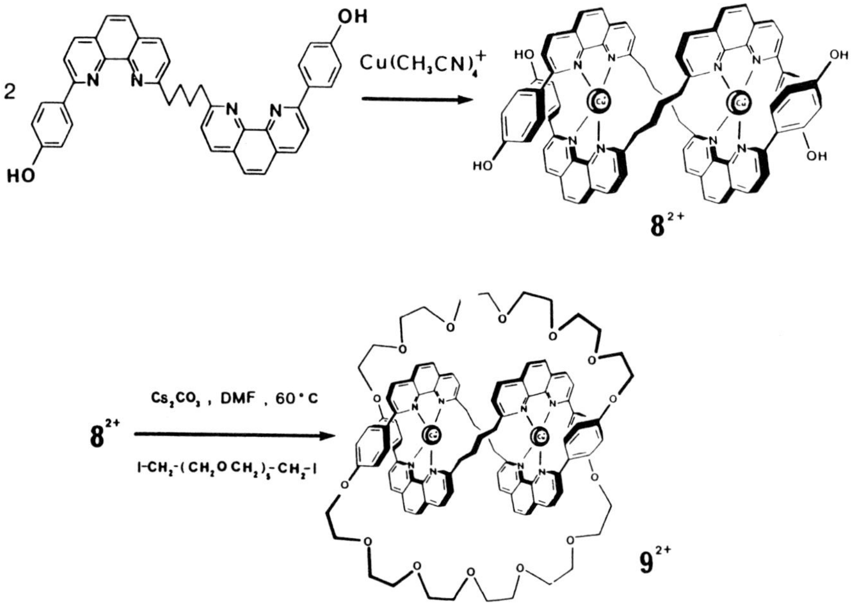

Figure 5

Demetallation of the bis-copper (I) complex leads to the free ligand **10** represented in Figure 6. It consists of a knotted 86-membered ring. This is a chiral system due to the topological chirality of a trefoil knot, contrary to the "normal" unknotted 86-membered ring. These two compounds (knotted and unknotted rings) are true topological stereoisomers.

Although the procedure has to be greatly improved for becoming preparative, it leads to the first example of a chemical knotted closed curve. Such particular topologies are particularly relevant to molecular biology. In Nature, duplex circular DNA can be converted to highly knotted forms, making the synthesis of knots an even more challenging problem to the chemist. These knotted and interlocked forms of DNA are involved in recombination and replication processes.

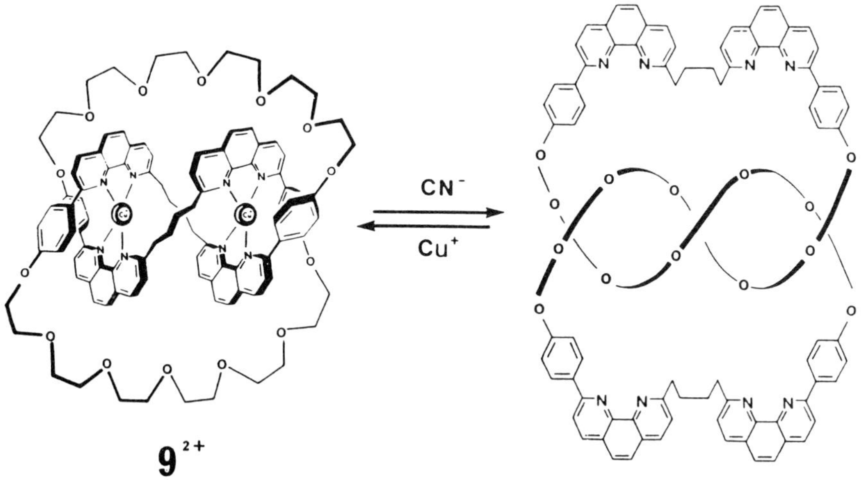

Figure 6

The X-ray structure of the dicopper (I) knotted complex 9^{2+} has very recently been solved.[10] It is in excellent agreement with the postulated structure of the molecule.

Acknowledgements

The CNRS is gratefully acknowledged for its constant financial support. I would also like to thank the many researchers who enthusiastically participated in this work, and in particular Dr. C.O. Dietrich-Buchecker.

REFERENCES

1. H.L. Frisch and E. Wasserman, J. Am. Chem. Soc., 1961, 83, 3789.
2. G. Schill, 'Catenanes, Rotaxanes and Knots', Academic Press, New York and London, 1971.
3. D.M. Walba, Tetrahedron, 1985, 41, 3161.
4. C.O. Dietrich-Buchecker, J.-P. Sauvage and J.-M. Kern, J. Am. Chem. Soc., 1984, 106, 3043.
5. C.O. Dietrich-Buchecker and J.-P. Sauvage, Chem. Rev., 1987, 87, 795 and references.

6. J.-P. Sauvage and J. Weiss, J. Am. Chem. Soc., 1985, 107, 6108.
7. C.O. Dietrich-Buchecker, A.K. Khemiss and J.-P. Sauvage, J. Chem. Soc., Chem. Comm., 1986, 1376.
8. C.O. Dietrich-Buchecker, C. Hemmert, A.K. Khemiss and J.-P. Sauvage, J. Am. Chem. Soc., in press.
9. C.O. Dietrich-Buchecker and J.-P. Sauvage, Angew. Chem., 1989, 101, 192.
10. C.O. Dietrich-Buchecker, J. Guilhem, C. Pascard and J.-P. Sauvage, submitted to Angew. Chem.

Subject Index

acetals	1,2
acetoacetic ester, derivative of	147
acetylation	145
acetylcholine	87
acid-catalyzed solvolysis	109
acyliminium intermediate	34
N-acyliminium intermediates	181
adenine	110,122
ajmalicine	29
akagerine	31, 36, 41
alcohol protecting groups	12
alkylation	144, 147
1, 3-allylic strain	199
π-allyltricarbonyliron lactone complex	7,10
Amanita phalloides	54
amanitin analogs	59
Amaryllidaceae	174
amatoxins	54, 55, 58
amino acid	54-56, 58, 63
γ-aminobutyric acid	87*ff*
9-amino-6, 7-dihydropyridopyrimidines	203
9-aminotetrahydropyridopyrimidines	197, 203
analgetic effect	187
anhydrolicorinium chloride	174
antiallergic	188,214
anticonvulsant agent	87
antimicrobial activity	103
Antirhea putaminosa	32
antirhine	31, 32, 35, 36
antithrombotic activities	187
antitumor activity	103, 122
anxiogenic	91
anxiolytic agent	87
apomorphine	160
aporphines	154-157
aprotic diazotization	158
aristolactams	154, 158, 159, 173
arylhydrazones	196

arylhydrazono derivatives	193
L-ascorbic acid, bound	68
ascorbigen,	68*ff*
acetyl derivatives	69
arachidonic acid and	83
ascorbic acid release	74
biological properties of	83
biological role of	83
CD-spectra of	73
derivatives as "shot-gun drugs"	81
HCl treatment of	72
immunomodulatory properties of	82
interaction with alkylamines	79
interaction with amines	78
interaction with arylalkylamines	79
interaction with arylamines	79
interferon and	84
interleukin 2 and	84
N-substituted	71
substituted	70
synthesis	69
transformation in acidic media	73
ascorbigen A	70
ascorbigen B	70
structure of	73
pseudo-*Aspidosperma*	46
asymmetrization	28, 37
asymmetry index, of BD	100
AT base pairs	109
AT binding selectivity	115
autofootprinting	105
avermectins	1,5,10,13
avicine	170
azadienes	169
azides	134, 136
barbiturates	88
2-benzenesulphonyl cyclic ethers	1,5
benzodiazepines	87*ff*
benzodiazepines,	
adrenal glands and	88
binding site of	88
CEF-curves of	96

Subject Index

conformers of	97*ff*
cyclic dimers of	101
dihedral angle in	98
ERC-MS spectra of	95
Fermi resonance in	99
gastric ulcer and	88
inversion	95
IR spectra of	99
linear associates of	99
MNDO calculation	98
monomers of	99
pseudo-boat conformation of	95
quantum chem. calc.	98
stereochemistry of	94
torsion angle in	100
X-ray of	95
benzodiazepine receptor,	88*ff*
affinity for	93
agonists	89
allosteric interaction of	88
antagonists	90
central	88
endogenous ligand of	89, 92
heterogeneity	88
inverse agonist	91
peripheral	88
partial antagonist	91
partial inverse agonist	91
benzyne-promoted dehydrogenation	163
binding constant	108
biological activity	54-56, 58, 63
Bischler-Napieralski cyclization	38, 156
bis-hydropyran spiroacetal	8
α-bromination	137
5′-bromoascorbigen	72
bromoalkylazetidinones	134
α-bromoaldoximes	136
5-bromoindole	70
9-bromopyridopyrimidine	203, 212
9-bromotetrahydropyridopyrimidine-3-carboxylate	201
α-bromo-O-silylaldoximes	137

cabbage	68
β-carbolines	89*ff*, 168, 169
carminomycin	82
catalytic activity	220, 222
catenand	223
catenand,	
electrochemical properties	224
kinetic properties	224
photophysical properties	224
topology of	224
catenane	223
catenate	223
CC-1065,	
interaction with DNA	103-105
X-ray structure of	104
CD spectra	64, 65
cleavamine, 20R-15,20-dihydro-	51
clavizepine	177
Cinchona	41
chirality, topological	226
cis stereochemistry	133
clay-fen	189, 213, 214
CNS-effects	188
corydaline	166
Corynanthe	32
Co(II) complex	220
Cotton effect	38, 63, 64, 66
CP 61405	8
crown-linked K^+ ions	145
Cu(I)-complex	224
cyclic ethers	1
cyclic ether sulphones	1,5
cyclic peptides	54, 55
cyclic urethanes	26
cycloaddition	139, 142
4+2 cycloadditions	130
cyclohexene	
cis, 1,2-dimethanol diacetate	29, 31, 37, 41
cis, 1,2-dimethanol monoacetate	32, 37, 38, 42
cis, 1,2-disubstituted	30
cysteine	54
cytotoxic activity	103, 112, 115, 118, 122, 126

Subject Index

daunorubicin	82
DBI, polypeptide	92
DBU	19, 20
decarbonylation	166
demetallation of catenates	223
desulphuration	55
diastereoselectivity	25
diazonium salts	189
dibromopyridopyrimidines	197
9, 9-dibromopyridopyrimidines	191
Diels-Alder reaction	130, 155
dieneamides	181
diepoxide	12
diesters	28
diethyl ethoxymethylenemalonate	189
dehydroaporphines	154, 155, 157
dihydropyran	1
1,4-dihydropyridine derivatives	144
di(indol-3-yl)methane	70
4,5-dioxoaporphines	154, 158
DISCOVER	46, 49
1,3-dipolar cycloadditions	130
DMAD	169
DMF-POCl$_3$ complex	204
DNA	58
DNA: CC-1065 interaction	104, 105
DNA,	
binding	109
binding affinity	106
covalent alkylation of	103-105, 109, 110, 118, 122
duplex circular	227
noncovalent binding	108
doxorubicin	82
double helix	226
dopamine	87
duguenaine	157, 158
dynamic NMR	95
pseudo-*Eburna*	46
Edman degradation	55, 56
electrochemical properties of catenands	224
emetine skeleton	169

enamidoester	181
enamine-imine equilibrium	24, 26
enol-imine tautomer	210
1-ethylindole	70
fagaronine	170
Fischer indolization	189, 193, 196
formyl-enamine tautomer	210
Fremy's salt	161
functionalized tetrahydropyrans	139
GABA	87*ff*
glucobrassicin	68
glycine	87
glycosidation	3,10
Hansch, method of	87
harman derivatives	92
heptapeptide	54
HLAD	219
H-nitrone	142
horse liver alcohol dehydrogenase(HLAD)	219
5-HT	87
hydrazones	193
9-hydrazonopyridopyrimidine	192, 193, 196
hydrogen bridge	24
hydrogen-deuterium exchange	189
hydropyran	8,10
hydroxylation	61
hydroxyl protecting groups	12
3-hydroxymethylindole	68, 70, 71
5-hydroxytryptamine	87
hypno-sedative agent	87
hypoxanthine	89
ICD, induced circular dichroism	106
indeno[1,2-b]pyridines	144
indeno[1,2-b]pyridines,	
acylation of	144
alkylation of	144, 147, 149, 151
catalytic hydrogenation of	150
hydrogenated	144

Subject Index

oxidation of	149
reactions of	144
reduction of	149
salts	149
indolopyrrolocarbazole	19
indoles	168
INOC (intramolecular nitrile oxide olefin cycloaddition)	135, 140, 141
inosine	89
interlaced systems	223
interlocked rings	223
Intermolecular Benzyne Cycloaddition (IBC)	155, 156, 158, 159, 163
intramolecular aryne cycloaddition	172, 173
intramolecular nucleophilic addition	176
ionophore antibiotic CP 61405	8
IOOC (intramolecular oxime olefin cycloaddition)	141
9-isonitrosopyridopyrimidine	212
ISOC (intramolecular silyl nitronate olefin cycloaddition)	140, 141
isoquinoline alkaloids	154
isoretronecanol	133
isoxazolidine	141
isoxazolines	130
isoxazolo pyrrolidine	141
isoxazolothiophenes	140
Japp-Klingeman reaction	189, 205
julolidine	181
kinetic properties of catenands	224
knot	223
knot, molecular trefoil	227
lactam	7
β-lactam	133
β-lactam triazolines	135
lactol	1
lactol ether	1
lactones	7
Lawesson's reagent	145
lupinin	181
lysicamine	166
macrolactonization	13
macrolide	10

86-membered ring (knotted)	226
meroquinene	31, 41, 46
1-methylindole	70
Michael addition	23, 142
milbemycins	1,5
minor grove binding	106, 108
mirosinase	68
MM2-calculations	137
monoterpenoid alkaloids	29
mustard oil glucoside	68
NBS	137, 190
neoglucobrassicin	68
Ni(I) complex	224
nitidine	170
nitrile oxide	130, 133, 136, 140
nitro olefins	139
non-conjugated formyl group	207
noncovalent binding affinity	106
noraporphines	156
norcepharadione B	158
norepinephrine	87
norphalloin	56
norpontevedrine	159
nuphar alkaloids	180,181
nupharidine	180
nupharopumiline	180
Nymphaeaceae	180
oleandrose	10
opiate receptor	88
oxidation of alcohols	8,13
oxime olefin cycloadditions	142
oximes	142
oxoaporphines	154, 161, 165
4-oxo-4H-pyrido(1,2-a)pyrimidine-3-carboxylates	189
7-oxoquinolizidine-4-one	184
pentacoordinated complex with $Zn(NO_3)_2$	222
peptides	54
peptides, hydrolysis of	55
phallotoxins, synthesis of	56

Subject Index

pharmacological properties of BD	87*ff*
9-phenyl-aminotetrahydropyridopyrimidinones	199
pheromone	181
photochemical electrocyclization	162
photocyclization	18, 21, 22
photooxidation	181
photophysical properties of catenands	224
picritoxin	88
Pictet-Spengler reaction	34
pig pancrease lipase	31, 38
piperidine	
cis-N-benzyloxycarbonyl-3,5-bismethoxycarbonyl	
(enzymatic hydrolysis of)	48
(half ester of)	48
(protection of)	49
dimethyl-*cis*-3,5-bismethoxycarbonyl	46, 47
cis-3,5-disubstituted	46
PLE	31, 37, 43
poly(indol-3-yl)methane	71
polymeric azide reagent	134
polypeptide DBI	92
pontevedrine	159
porcine liver esterase	31, 37, 43
PPL	31, 38
Pschorr cyclization	154
printing	105
proconvulsant agent	91
protecting groups	12
protoberberine alkaloids	166-169
pumiliotoxine	22
4H-pyrido(1,2-a)pyrimidin-4-one	187
3H-pyrimidin-4-one	191
pyrrolizidines	143
quinolizidine	34
quinolizidine derivatives	180,181
quinolizidines	143
4-quinolizidinone	180
radical cyclization	124
radioligand assay	97
reductive amination	22
regioselective halogenation	191

retro-Michael reaction	147
ring contraction	214
RNA	58, 59
rutaecarpine alkaloids	195
selenenylation	13
silyl enol ether	3, 13
silyl nitronate	140
spiroacetal	1, 5, 8
spiro cyclization	24
spiropiperidine	22, 26
staurosporine	18
Stephania suberosa	167
stereoisomers, topological	226
stereoselective synthesis	142, 143
stereoselectivity	26, 141
stereospecific	142
stereospecificity	140
stereospecific transformations	135
Streptomyces zelensis	103
Strychnos usambariensis	36
styrylamides	174
sulfenyl chlorides	56, 57
suriclon	89
σ-symmetrical diesters	28
symmetrization-asymmetrization	29
template effect, three dimensional	223
tetrahydropyridopyrimidinones	189, 191
6, 7, 8, 9-tetrahydropyridopyrimidinones	188
1, 6, 7, 8-tetrahydropyrimidinones	198
tetrapropylammonium perruthenate (TPAP)	8, 12, 13
topological chirality	226
topological stereoisomers	226
topology of catenands	223, 224
toxins,	
identification of	60
spectral properties	61
TPAP	8, 12, 13
transition metal	223
tricarbonyliron lactone complex	7, 10
trichoguattine	162

Subject Index

tryptamine	195
tryptathionine	54-56
tryptathionine,	
chirality of	60
cleaved by reduction	56
oxidation of	54
tryptophan	54
tryptophan, oxidation of	58
unsaturated oximes	139
urethane protecting group for nitrogen	39
Vilsmeier-Haack formylation	189, 206
vinylazetidine	137
2-vinylazetidinones	133, 136
vinyl nitroso species	137
virosa mushrooms	54
virotoxins	59, 60
vitamin C	68
Wittig reaction	206
xanthenes	176
X-ray structure of trefoil knot	227
zinc bromide	3, 10
zopiclon	89
Zn(II) complex,	
dimeric	219, 220
monomeric	221
NMR investigations of monomeric	221
oligomeric	219